TECHNIQUES FOR TRACKING, EVALUATING, AND REPORTING THE IMPLEMENTATION OF NONPOINT SOURCE CONTROL MEASURES

III. URBAN

Final
January 2001

Prepared for

Nonpoint Source Pollution Control Branch
Office of Water
United States Environmental Protection Agency
Washington, DC

Prepared under

EPA Contract No. 68-C-99-249
Work Assignment No. 0-25

CONTENTS

List of Tables

List of Figures

CHAPTER 1. INTRODUCTION

1.1 Purpose of Guidance

This guidance focuses on the methods that can be used to inventory specific types of urban BMPs and the design of monitoring programs to assess implementation of urban management measures and BMPs, with particular emphasis on statistical considerations.

This guidance is intended to assist federal, state, regional, and local environmental professionals in tracking the implementation of best management practices (BMPs) used to control urban nonpoint source pollution. Information is provided on methods for inventorying BMPs, the design and execution of sampling programs, and the evaluation and presentation of results. The more regulated and stable nature of urban areas present opportunities for inventorying all BMPs versus the statistical sampling required to assess BMP implementation for agriculture or forestry. Inventorying BMP implementation requires establishing a program that tracks the implementation or operation and maintenance of all BMPs of certain types (e.g., septic tanks and erosion and sediment control practices). The guidance can help state and local governments by providing a subset of controls, both structural and nonstructural, that can be sampled for:

- inspection programs,
- maintenance oversight, and
- implementation confirmation.

The focus of chapters 3 and 4 is on the statistical approaches needed to properly collect and analyze data that are accurate and defensible. A properly designed BMP implementation monitoring program can save both time and money. For example, the cost to determine the degree to which pollution prevention activities are conducted by an entire urban population would easily exceed most budgets, and thus statistical sampling of a subset of the population is needed. Guidance is provided on sampling representative BMPs to yield summary statistics at a fraction of the cost of a comprehensive inventory.

While it is not the focus of this guidance, some nonpoint source projects and programs combine BMP implementation monitoring with water quality monitoring to evaluate the effectiveness of BMPs in protecting water quality on a watershed scale (Meals, 1988; Rashin et al., 1994; USEPA, 1993b). For this type of monitoring to be successful, the scale of the project should be small (e.g., a watershed of a few hundred to a few thousand acres). Accurate records of all the sources of pollutants of concern, how these sources are changing (e.g., new development), and an inventory of how all BMPs are operating are vital for this type of monitoring. Otherwise, it is impossible to accurately correlate BMP implementation with changes in stream water quality. This guidance does not address monitoring the implementation and effectiveness of individual BMPs. It does provide information to help program managers gather statistically valid information to assess implementation of BMPs on a more general (e.g., statewide) basis. The benefits of implementation monitoring are presented in Section 1.3.

1.2 BACKGROUND

Because of the past and current successes in controlling point sources, pollution from nonpoint sources—sediment deposition, erosion, contaminated runoff, hydrologic modifications that degrade water quality, and other diffuse sources of water pollution—is now the largest cause of water quality impairment in the United States (USEPA, 1995). Recognizing the importance of nonpoint sources, Congress passed the Coastal Zone Act Reauthorization Amendments of 1990 (CZARA) to help address nonpoint source pollution in coastal waters. CZARA provides that each state with an approved coastal zone management program develop and submit to the U.S. Environmental Protection Agency (EPA) and National Oceanic and Atmospheric Administration (NOAA) a Coastal Nonpoint Pollution Control Program (CNPCP). State programs must "provide for the implementation" of management measures in conformity with the EPA *Guidance Specifying Management Measures For Sources Of Nonpoint Pollution In Coastal Waters*, developed pursuant to Section 6217(g) of CZARA (USEPA, 1993a). Management measures (MMs), as defined in CZARA, are economically achievable measures to control the addition of pollutants to coastal waters, which reflect the greatest degree of pollutant reduction achievable through the best available nonpoint pollution control practices, technologies, processes, siting criteria, operating methods, or other alternatives (all of which are often referred to as BMPs). Many of EPA's MMs are combinations of BMPs. For example, depending on site characteristics, implementation of the Construction Site Erosion and Sediment Control MM might use the following BMPs: brush barriers, filter strips, silt fencing, vegetated channels, and inlet protection.

CZARA does not specifically require that states monitor the implementation of MMs and BMPs as part of their CNPCPs. State CNPCPs must however, provide for technical assistance to local governments and the public for implementing the MMs and BMPs. Section 6217(b) states:

> Each State program . . . shall provide for the implementation, at a minimum, of management measures . . . and shall also contain . . . (4) The provision of technical and other assistance to local governments and the public for implementing the measures . . . which may include assistance . . . to predict and assess the effectiveness of such measures . . . .

EPA and NOAA also have some responsibility under Section 6217 for providing technical assistance to implement state CNPCPs. Section 6217(d), Technical assistance, states:

> [NOAA and EPA] shall provide technical assistance . . . in developing and implementing programs. Such assistance shall include: . . . (4) methods to predict and assess the effects of coastal land use management measures on coastal water quality and designated uses.

This guidance document was developed to provide the technical assistance described in CZARA Sections 6217(b)(4) and 6217(d), but the techniques can be used for similar programs and projects. For instance, monitoring projects funded under Clean Water Act (CWA) Section 319(h) grants, efforts to implement total maximum daily loads developed under CWA Section 303(d), stormwater permitting programs, and other programs could all benefit from knowledge of BMP implementation.

Methods to assess the implementation of MMs and BMPs, then, are a key focus of the technical assistance to be provided by EPA and NOAA. Implementation assessments can be done on several scales. Site-specific assessments can be used to assess individual BMPs or MMs, and watershed assessments can be used to look at the cumulative effects of implementing multiple MMs. With regard to "site-specific" assessments, individual BMPs must be assessed at the appropriate scale for the BMP of interest. For example, to assess the implementation of MMs and BMPs for erosion and sediment control (E&SC) at a construction site, only the structures, areas, and practices implemented specifically for E&SC (eg., protection of natural vegetation, sediment basins, or soil stabilization practices) would need to be inspected. In this instance the area physically disturbed by construction activities and the upslope area would be the appropriate site and scale.

However, if a state without a centralized E&SC program were assessing erosion and E&SC in an area (e.g., coastal) of concern, it might assess municipal E&SC programs. In this instance the "site" would be each urban area and implementation of municipal regulations, inspection and enforcement programs, etc. would be checked. For bridge runoff management, the scale might be bridges over waterways that carry and average daily traffic of 500 or more vehicles and the sites would be individual bridges that meet this requirement. Site-specific measurements can then be used to extrapolate to a program, watershed, or statewide assessment. There are instances where a complete inventory of MM and BMP implementation across an entire watershed or geographic area is preferred.

1.3 Types of Monitoring

The term *monitor* is defined as "to check or evaluate something on a constant or regular basis" (Academic Press, 1992). It is possible to distinguish among various types of monitoring. Two types, implementation at a specific time (i.e., a snapshot) and trend (i.e., trends in implementation) monitoring, are the focus of this guidance. These types of monitoring can be used to address the following goals:

- Determine the extent to which MMs and BMPs are implemented in accordance with relevant standards and specifications.

- Determine whether there has been a change in the extent to which MMs and BMPs are being implemented.

In general, implementation monitoring is used to determine whether goals, objectives, standards, and management practices are being implemented as detailed in implementation plans. In the context of BMPs within state CNPCPs, implementation monitoring is used to determine the degree to which MMs and BMPs required or recommended by the CNPCPs are being implemented. If CNPCPs call for voluntary implementation of MMs and BMPs, implementation monitoring can be used to determine the success of the voluntary program (1) within a given monitoring period (e.g., 1 or 2 years); (2) during several monitoring periods, to determine any temporal trends in BMP implementation; or (3) in various regions of the state.

Trend monitoring involves long-term monitoring of changes in one or more parameters. As discussed in this guidance, public attitudes, land use, and the use of various urban BMPs are examples of parameters that could be measured with trend monitoring. Isolating the impacts of MMs and BMPs on water quality requires trend monitoring.

Because trend monitoring involves measuring a change (or lack thereof) in some parameter over time, it is necessarily of longer duration and requires that a baseline, or starting point, be established. Any changes in the measured parameter are then detected in reference to the baseline.

Implementation and the related trend monitoring can be used to determine (1) which MMs and BMPs are being implemented, (2) whether MMs and BMPs are being implemented as designed, and (3) the need for increased efforts to promote or induce use of MMs and BMPs. Data from implementation monitoring, used in combination with other types of data, can be useful in meeting a variety of other objectives, including the following (Hook et al., 1991; IDDHW, 1993; Schultz, 1992):

- To evaluate BMP effectiveness for protecting natural resources.
- To identify areas in need of further investigation.
- To establish a reference point of overall compliance with BMPs.
- To determine whether landowners are aware of BMPs.
- To determine whether landowners are using the advice of urban BMP experts.
- To identify any BMP implementation problems specific to a land ownership or use category.
- To evaluate whether any urban BMPs cause environmental damage.
- To compare the effectiveness of alternative BMPs.

MacDonald et al. (1991) describes additional types of monitoring, including effectiveness, baseline, project, validation, and compliance monitoring. As emphasized by MacDonald and others, these monitoring types are not mutually exclusive and the distinctions among them are usually determined by their purpose.

Effectiveness monitoring is used to determine whether MMs or BMPs, as designed and implemented, are meeting management goals and objectives. Effectiveness monitoring is a logical follow-up to implementation monitoring, because it is essential that effectiveness monitoring include an assessment of the adequacy of the design and installation of MMs and BMPs. For example, the objective of effectiveness monitoring could be to evaluate the effectiveness of MMs and BMPs *as designed and installed*, or to evaluate the effectiveness of MMs and BMPs *that are designed and installed adequately or to standards and specifications*. Effectiveness monitoring is not addressed in this guide, but is the subject of another EPA guidance document, *Monitoring Guidance for*

Determining the Effectiveness of Nonpoint Source Controls (USEPA, 1997).

1.4 Quality Assurance and Quality Control

An integral part of the design phase of any nonpoint source pollution monitoring project is quality assurance and quality control (QA/QC). Development of a quality assurance project plan (QAPP) is the first step of incorporating QA/QC into a monitoring project. The QAPP is a critical document for the data collection effort inasmuch as it integrates the technical and quality aspects of the planning, implementation, and assessment phases of the project. The QAPP documents how QA/QC elements will be implemented. It contains statements about the expectations and requirements of those for whom the data is being collected (i.e., the decision maker) and provides details on project-specific data collection and data management procedures designed to ensure that these requirements are met. Development and implementation of a QA/QC program, including preparation of a QAPP, can require up to 10 to 20 percent of project resources (Cross-Smiecinski and Stetzenback, 1994). A thorough discussion of QA/QC is provided in Chapter 5 of EPA's *Monitoring Guidance for Determining the Effectiveness of Nonpoint Source Controls* (USEPA, 1997).

1.5 Data Management

Data management is a key component of a successful MM or BMP implementation monitoring effort. The system used—including the quality control and quality assurance aspects of data handling, how and where data are stored, and who manages the stored data—determines the reliability, longevity, and accessibility of the data. Provided that data collection was well planned and executed, an organized and efficient data management system will ensure that the data:

- Can be used with confidence by those who must make decisions based upon it,
- Will be useful as a baseline for similar data collection efforts in the future,
- Will not become obsolete quickly, and
- Will be available to a variety of users for myriad applications.

Serious consideration is often not given to a data management system prior to a collection effort, which is precisely why it is so important to recognize the long-term value of a small investment of time and money in proper data management. Data management competes with other priorities for money, staff, and time, and if the importance and long-term value of data management is recognized early, the more likely it will be to receive sufficient funding. Overall, data management might account for only a small portion of a project's total budget, but the return is great, considering that data can be rendered virtually useless if data are not managed adequately.

Two important aspects of data that should be considered when planning the initial collection and management systems are data life cycle and accessibility. The cycle has 5 stages: (1) data are collected; (2) data are checked for quality; (3) data are entered into a database; (4) data are used, and (5) data eventually become obsolete. The expected usefulness and life span of the data should be considered during the initial stages of planning a data collection effort, when the money, staff, and time devoted to data collection must be weighed against its usefulness and longevity. Data with limited uses and likely to become obsolete soon after collection are a poorer investment decision than data with multiple applications and long life spans.

Data accessibility is a critical factor in determining its usefulness. Data attains highest value if widely accessible, if access requires the least staff effort, and if used by others conveniently. If date are stored where obtainable (with little assistance), use and sharing are more likely. The format for data storage determines how conveniently data can be used. Electronic storage in widely available and used data formats makes for convenience. Storage as only a paper copy buried in a report, where any analysis requires entry into an electronic format or time-consuming manipulation, makes data extremely inconvenient and thus unlikely to be used.

The following should be considered for the development of a data management strategy:

- *What level of quality control should the data be subject to?* Data to be used for a variety of purposes or for important decisions merit careful quality control checks.

- *Where and how will the data be stored?* The options for data storage range from a printed final report on a bookshelf to an electronic data base accessible to government agencies and the public. Determining where and how data will be stored requires careful consideration of the question: *How accessible should the data be?*

- *Who will maintain the data base?* Data stored in a large data base might be managed by a professional manager, while data kept in agency files might be managed by people of various backgrounds over time.

- *How much will data management cost?* As with all other aspects of data collection, data management costs money and must be balanced with all other project costs.

CHAPTER 2. METHODS TO INVENTORY BMP IMPLEMENTATION

Because the potential for serious water quality degradation is high in urban areas, it is important to have a means to track the implementation of BMPs used to control urban nonpoint source pollution and a means to measure what is being done to address it. The activities in urban areas that generate polluted runoff are usually concentrated in a small area, discharging to only one or two water bodies, and diverse, contributing a variety of pollutants. Although programs exist for statewide tracking of BMPs for forestry and agriculture (see Adams, 1994; Delaware DNREC, 1996) and some studies of BMP implementation in urban areas have been done (see Pensyl and Clement, 1987), comprehensive urban-area BMP tracking programs are still not the norm. In some ways, tracking BMPs in urban areas can be easier than tracking those for forestry or agriculture. For instance, once an area is developed and structural BMPs are installed, there is little change unless problems require retrofits. If an inventory of BMPs (e.g., stormwater ponds, swales, buffer strips) is done, the information can be stored in a database and used for a variety of purposes. Also, many of the urban pollutant-generating activities are permitted (e.g., construction) or regulated in some other manner (e.g., septic tank operation and maintenance), providing a paper trail of information. These advantages can result in a more complete assessment of urban BMP implementation. In some instances it is possible to inventory and track over time the implementation status of all BMPs of certain types. For those urban areas that have not compiled existing codes, regulations, and permitting requirements, it is recommended that an inventory be created.

2.1 Regulated Activities

To regulate urban NPS water pollution, states employ a variety of legal mechanisms, including nuisance prohibitions, general water pollution discharge prohibitions, land use planning and regulation laws, building codes, health regulations, and criminal laws (Environmental Law Institute, 1997). Many states delegate some of these authorities to units of local government or conservation districts. Although not all pollutant-generating activities are covered by these mechanisms, the applicable mechanisms present opportunities for inventorying BMP implementation. The urban activities that are regulated in some manner include erosion and sediment control, onsite sewage disposal systems (septic tanks), runoff from development sites, construction, and site-specific activities (e.g., oil and grit separators at gas stations). Perhaps the best mechanism for collecting information for tracking BMP implementation is requiring permits for certain activities. A permitting system places on the applicant the burden of obtaining and supplying all necessary data and information needed to get the permit. Two types of permits are generally issued—construction and operating. Issuance of these permits encourages construction and operation of BMPs in compliance with local laws and regulations.

2.1.1 Erosion and Sediment Control

Most urban areas have laws requiring the control of sediment erosion at construction sites. These laws are usually implemented as part of the building permit process. The material required as part of the building permit process (including site clearing plans, drainage plans, landscaping plans, and erosion and sediment control plans) can provide a wealth of information on proposed BMP implementation. Because site clearing and building activities occur in a short time period, tracking of implementation of BMPs for erosion and sediment control should be done on a real-time basis.

Many states and municipalities employ inspectors to monitor BMP implementation at building sites. Site inspections are critical to determining actual BMP implementation. Paterson (1994), in a survey of construction practices in North Carolina, found that nearly 25 percent of commonly prescribed construction BMPs (e.g., storm drain inlet protection and silt fences) that were included in erosion and sediment control plans had not actually been implemented (see ***Example 1***). Employing adequate numbers of site inspectors can be expensive. To counter a lack of BMP implementation and to overcome a shortage of construction site inspection staff, the state of Delaware developed an effective program to monitor implementation of BMPs

This investigation looked at more than 1,000 construction practices that had been included in 128 erosion and sediment control plans in nine North Carolina jurisdictions. The nine jurisdictions were selected to be representative of North Carolina's three physiographic regions (coastal plain, piedmont, and mountain) across three levels of program administration (municipal, county, and state). Project sites were randomly selected from lists of permitted construction projects provided by each jurisdiction . The implementation of erosion and sediment control practices was evaluated in terms of whether the practices had been installed adequately and whether they were being maintained adequately.

The survey provided information on the following aspects of erosion and sediment control practices:

- Which practices administrators thought were useful practices and which they thought were poor performers.
- What administrators thought were the causes of practice failure (e.g., poor installation, poor maintenance).
- The number of construction practices never installed even though they were on the erosion and sediment control plan.
- Which practices were poorly installed/constructed/maintained and the installation/construction/maintenance problem.
- Which practices were prescribed in erosion and sediment control plans.
- Which recommended practices performed worse than less-favored practices.
- What problems were associated with installation of the practices.

The investigators determined that the major problems associated with installation were a lack of suitable training to install the erosion and sediment control practices properly and vagueness in the erosion and sediment control plan concerning installation specifications. The major problems associated with maintenance were neglect of the practices after installation and initial design flaws.

Example 1 . . . *Review of erosion and sediment control plans in North Carolina.* (Paterson, 1994) for erosion and sediment control at construction sites (Center for Watershed Protection, 1997) (see ***Example 2***).

2.1.2 Septic Systems

Cesspools, failed septic systems, and high densities of septic systems can contribute to the closure of swimming beaches and shellfish beds, contaminate drinking water supplies, and cause eutrophication of ponds and coastal embayments. Onsite sewage disposal systems (OSDS) are usually locally regulated by building codes and health officials (Environmental Law Institute, 1997). A variety of permit requirements are used to regulate their siting, installation, and operation and maintenance.

Several innovative programs have been developed to track implementation of BMPs for OSDS (see ***Examples 3 and 4***). Program

The state of Delaware's program requires some builders to hire independent inspectors, who are officially known as construction reviewers. These reviewers monitor implementation of erosion and sediment control BMPs at selected construction sites.

The construction reviewers are certified and periodically recertified in erosion and sediment control by the state of Delaware and provide onsite technical assistance to contractors. They are required to visit sites at least weekly and to report violations and inadequacies to the developer, contractor, and erosion and sediment control agency. Their reports are reviewed by government inspectors. Local or state erosion and sediment control agencies are still responsible for spot checking sites and issuing fines or other penalties. Reviewers can lose their certification if spot checks reveal that violations were not reported. Since its inception in 1991, 340 people have been certified as construction reviewers.

Successful implementation of a program similar to Delaware's would require tailoring it to regional circumstances and conditions. Key aspects of the program in Delaware include the following:

- Full-time staff were assigned to administer the program.
- Criteria for selection of appropriate sites for the use of construction reviewers were established.
- A training program and certification course were developed to support the program.
- Reporting criteria were specified.
- Oversight by a professional engineer was incorporated.
- Specific spot check scheduling was determined.
- Recourse for fraudulent inspection results was incorporated.
- Enforcement actions for contractors who violate erosion and sediment control plans were included.
- The program was piloted in a test area.
- Objective monitoring criteria were developed to evaluate the program.
- A process for revision to the program based on performance was included.

Example 2 . . . *Delaware's construction reviewer program.* (CWP, 1997)

In the Buzzards Bay area a need to track information related to OSDS permitting and inspection and maintenance was identified. Municipal boards of health in this area are responsible for implementing and overseeing state regulations for OSDS. The boards of health lacked the ability to efficiently and effectively monitor permits and inspection and maintenance information, due to insufficient staffing and information-processing equipment and systems. They had been overburdened with processing new permits, with the result that tracking past permits and past orders of noncompliance and reviewing pump-out reports were tasks often left undone.

Project: The SepTrack Demonstration Project provided computers and specialized software to communities fringing Buzzards Bay to enable them to better manage information related to onsite septic systems. This helped to identify patterns of septic system failure and freed staff time for better design review and enforcement.

Project Goal: To better enable each board of health to track septic system permits and inspection and maintenance information by reducing information management and retrieval burdens on boards of health, thereby allowing time to enhance protection of public health and the environment.

Accomplishments: Computers and specialized software were provided to 11 boards of health in the Buzzards Bay watershed. Funding was provided to transfer old permit information and septic pumping records in each community into the SepTrack database. The project was welcomed with enthusiasm by most municipalities, and many communities outside the demonstration area have requested copies of the SepTrack software.

Most boards of health receive monthly reports from sewage treatment plants with information on pumpouts provided by septage haulers. In Massachusetts, the haulers must report the source of their septage. Frequent pumping at a property is often a sign of a failing septic system. With SepTrack, a list of frequently pumped systems is provided automatically. In one town, this listing highlighted a town-owned property as one with a failing system and revealed inconsistencies in septage hauler information. In another town, public works water and sewer information in the SepTrack system revealed that 200 homes along an embayment had never been connected to a sewer line. The board of health required that this neighborhood connect to the existing sewer.

Example 3 . . . *Buzzards' Bay SepTrack System.* (USEPA, undated)

features that provide data that can be used to track implementation include the following:

- Building codes with design, construction, depth to water table, and soil percolation standards.
- Permitting of systems.
- Periodic inspections for compliance including whenever the system is pumped, the property is sold, or a complaint is filed.
- Requirements that the system be pumped periodically or if the property is sold and that the septage hauler file a report with the local health department.
- Dye testing of systems in areas of concern.

Marin County, California, biennial onsite system inspection program. Marin County, California, modified its code and established the requirement for a county-administered biennial onsite system inspection program. Part of the inspection program is a Certificate of Inspection, issued when the system is built and renewed every 2 years. Every 2 years a letter is sent to inform the property owner that an inspection is required. The owner must schedule an inspection and pay a renewal fee. Homeowners have the option of having the inspection performed by a county-licensed septic tank pumper with supervision by a county field inspector. Should repair or pumping be required, the homeowner must submit proof of repair or pumping before the certificate is renewed. New certificates are valid regardless of any change in home ownership prior to the certificate expiration date. The Certificate of Inspection must be valid and current when home ownership is transferred (Roy F. Weston, 1979 (draft)).

Wisconsin onsite wastewater treatment system installer certification. Wisconsin requires that onsite wastewater treatment system installers be certified by the state as either Plumbers or Restricted Sewer Plumbers. In addition, the state recently replaced the percolation test with a site-specific soil, drainage, and morphological evaluation that must be performed by a Certified Soil Tester.

Allen County, Ohio, Department of Health Monitoring Program. The Department of Public Health in Allen County, Ohio, monitors approximately 3,000 onsite disposal systems. Important components of the monitoring program include the following:

- Maintenance of a computerized billing process and paper files of inspection results and schedules.
- Permit issuance for all new systems. Afterward, the annual billing serves as the permit.
- Annual inspection of all aerobic systems covered under the program.
- Notification sent to property owners in advance of inspections.
- Inspections for loan certifications. Inspection is free for systems covered under the permit program.

Combination visual and chemical monitoring program, Santa Cruz, California. The San Lorenzo River watershed in Santa Cruz County, California, is encompassed by the Santa Cruz wastewater management zone. The wastewater management zone monitors the systems in the watershed as follows (Washington State, 1996):

- Maintaining a database with information on system ownership and locations, permits, loan certifications, complaints, failure and inspection results, and schedules.
- Assigning to each system a classification that determines the operations requirements, fee schedules, inspection frequencies, and property restrictions.
- Conducting initial inspections for all systems to assess system condition.
- Inspecting systems that meet standard requirements every 6 years; inspecting other systems every 1 to 3 years. (The health agency performs all inspections. Property owners are not notified of upcoming inspections).
- Administering public education programs through direct outreach and distribution of brochures.
- Monitoring surface water quality for fecal coliform and nitrate.

Example 4 . . . *Tracking onsite sewage disposal systems.*

2.1.3 Runoff Control and Treatment

It is possible to inventory and track all structural BMPs in a given geographic area over time. Such a project requires a large effort and has been used only when a state or watershed (e.g., Chesapeake Bay basin) is trying to reach a specific water quality goal. Such efforts may become more common in the future as states implement the Clean Water Act Section 303(d) total maximum daily load (TMDL) program for impaired waters. For example, an entire 94-square mile area of the Anacostia River watershed in Prince George's County, Maryland, was inventoried to

- Identify and document water resource problem areas and potential retrofit sites.
- Evaluated existing stormwater management facilities from water quality and habitat enhancement perspectives.
- Make recommendations for retrofit.
- Present information derived in a format useful to public agency personnel.

The investigators collected information on contributory drainage area, land ownership, land use/zoning, soils, areas of ecological or scenic significance, presence of wetland areas, storm drain outfall size and location, storm water management facility design specifications, ownership, maintenance responsibilities, base flow conditions, stream channel condition, and canopy coverage and riparian habitat conditions. The information was used to make management decisions on BMP retrofits, stream restoration, and installation of new BMPs. Another example of a state inventorying BMPs for the control of urban runoff is presented in ***Example 5***.

2.2 Tracking BMP Operation and Maintenance

In many instances the extent of proper operation and maintenance of a BMP is as important as the proper design and installation of the BMP. Regular inspection of BMP operation and maintenance can provide an indication of how a nonpoint source control program is advancing. Such inspections can also identify BMPs that need repairs or retrofits as well as identify areas that require additional management resources. If the right types of information are collected when a BMP is installed, the task of tracking operation and maintenance as well as ascertaining or monitoring effectiveness is much easier. BMP operation and maintenance can also be tracked through review of the BMP maintenance backlog. A large maintenance backlog indicates that additional resources are required to ensure proper operation.

Many of the examples presented earlier in this chapter contain information on how BMP operation and maintenance was tracked by the responsible agency. Lindsey et al. (1992) investigated the functioning and maintenance of 250 storm water BMPs in four Maryland counties and documented a need for improved inspection and maintenance. They found excessive sediment and debris in many devices and growth of woody or excessive vegetation and the need for stabilization near many. These problems had led to one-quarter of all basins (infiltration, wet, and dry) having lost more than ten percent of their volume and eroding embankments at more than one-third of all facilities. The BMPs were assessed as to the following maintenance criteria:

A comprehensive survey of infiltration devices was conducted in the state of Maryland to quantify the installation of the devices during the first 2 years after enactment of the Stormwater Management Act in that state (Pensyl and Clement, 1987). During the survey, state agency personnel, in cooperation with local county agencies, collected the data through actual site inspections. A separate inspection form was completed for each site inspection.

The following information was obtained during each site inspection:

- The type of infiltration device in use.
- The number of infiltration devices in use.
- The means of entry of runoff into the infiltration device.
- Whether the infiltration device was functioning.

The data were compiled by county to determine the following:

- The types of infiltration devices in use.
- The total number of infiltration devices installed.
- The total number of infiltration devices in each county.
- The total number of functioning infiltration devices per county.
- The total number of each type of infiltration device per county.
- The total number of each type of infiltration device that was functioning and nonfunctioning in each county.
- The percentage of functioning infiltration devices in each county.

The data were compiled by infiltration device to determine the following:

- The total number of each type of infiltration device in the survey area.
- The number and percentage of functioning and nonfunctioning infiltration devices of each type.
- Whether functioning infiltration devices were associated with buffer strips and drainage area stabilization.
- Whether functioning infiltration devices had obvious sediment entry, needed maintenance, or had standing water.

RESULTS

From the site inspection survey, the following was determined:

- The number of infiltration devices installed in the state.
- The number and percent of functioning infiltration devices.
- The type of infiltration device with the greatest percent of those installed that were functioning.
- The overall success rate of infiltration devices in the state (i.e., 67%).
- Which infiltration practices have a low success rate.
- The likely reasons for the failure of infiltration devices to function properly.
- Recommendations to improve the state storm water management program.

Example 5 . . . *Maryland survey of infiltration devices.*

- Facility functioning as designed.
- Quantity controlled as designed.
- Quality benefits produced by facility.
- Enforcement action needed.
- Maintenance action needed.

Several models were used to analyze the results of the field study, and the inspectors found that the conditions of the different types of BMPs varied significantly.

2.3 Geographic Information Systems and BMP Implementation/ Effectiveness

Geographic information systems (GIS) are useful for characterizing the features of watersheds in the form of spatial relationships in a manner that permits gaining much more information from data than could be obtained from them in the form of separate, unrelated databases. Spatial relationships among the locations of pollution sources, land uses, water quality data, trends in population and development, infrastructure, climatological data, soil type and geological features, and any other data that can be represented graphically and might be perceived as related to BMP implementation and water quality management can be incorporated into a GIS. In addition, nongraphical data can be incorporated into a GIS so they can be analyzed with respect to the graphical data. Nongraphical data include such things as dates of inspections and BMP maintenance, types of materials used in infrastructure, sizes of pipes and storm water inlets, and so forth.

A computer interface between a database, a GIS, and a storm water model was created for Jefferson Parish, Louisiana, to develop a computer simulation model for studying storm water runoff events, planning future capital drainage projects, and developing alternative management scenarios (Barbé et al., 1993).

The following graphical information was stored in the GIS: 1-foot contours, sidewalks, building outlines, aboveground and belowground public and private utilities, fences, water features, vegetation, parcels, political boundaries, and soil types. Nongraphical data on sewers and storm drainage were also stored for reference: pipeline size; pipe construction material; location of pipelines; and location, material, and depth of manholes. Similar information on streets was incorporated.

Robinson and Ragan (1993) note that the CWA Section 319 requirements—i.e., to submit reports that detail the amount of navigable waters affected by nonpoint sources, the types of nonpoint source affecting water quality, and the BMP program designed to control them—will require local governments

Robinson and Ragan (1993) correlated a nonpoint source model developed by the Northern Virginia Planning District Commission with mapping coordinates to determine the spatial distribution of nonpoint source constituents. The nonpoint source model approximates loading rates of several nonpoint source constituents from a relationship between land use and soil type. Robinson and Ragan developed the GIS themselves rather than using a vendor-sold system because the custom-made GIS was easier to use and did not require specialized training or modification of a standard GIS for their particular application.

Loudoun County, Virginia, grew enormously from 1980 to 1990. The primary source of drinking water is ground water obtained through wells and springs, so the county enacted regulations to require hydrogeologic studies to support proposals for new rural subdivisions. The county developed a comprehensive environmental GIS that incorporates a ground water database with information on water well yields, well depth, depth to bedrock, storage coefficients, underground storage tanks, landfills, sewage disposal systems, illegal dump sites, sludge application sites, and chemical analyses of ground water. The ground water database is linked to environmental mapping units (e.g., bedrock) to generate information such as the distribution of geology and well yields; the density, type, and status of potential pollution sources; and ground water quality as it relates to land use and geology (Cooper and Carson, 1993).

to integrate information on a regional basis and relate it through nonpoint source modeling in order to manage the quantity of data necessary to achieve the desired results and to conduct the simulations needed to support the decision-making process.

Data sets will have to be updated periodically, particularly with respect to land use, infrastructure, population, and demographics in developing areas. Using a GIS interfaced with nonpoint source pollution models is a good approach to achieve these ends (Robinson and Ragan, 1993). For example, EPA's Better Assessment Science Integrating Point and Nonpoint Sources (BASINS) is a system that integrates a GIS, national watershed data, and environmental assessment and modeling tools into one package. It allows users to add customized data layers, such as BMP implementation information, to existing data.

A GIS can be an extremely useful tool for BMP tracking since it can be used to keep track of and detect trends in BMP implementation, land treatment (e.g., areas of high use of fertilizer or pesticides), changes in land use (e.g., development), and virtually any data related to BMPs and water quality. An advantage of using a GIS for BMP tracking is the ability to update information and integrate it with existing data in a timely manner. Data are thereby made extremely accessible. Through the ability to correlate numerous types of data with a GIS, changes observed in data are more easily recognized. This permits managers to analyze the changes in one set of conditions with respect to other existing conditions within a particular geographical area and to arrive at plausible explanations, eliminate unplausible ones, and potentially to predict future problems.

GIS can be used as the basis for sampling for BMP tracking studies. Criteria for sampling can be chosen—for instance, age of BMP or elapsed time since the last inspection—and any BMPs that fail to meet the criteria can easily be eliminated from consideration. With all relevant information on BMPs in a single GIS, selection criteria for unrelated characteristics (e.g., retention capacity and most recent inspection date) can be correlated easily to arrive at a subset that meets all of the desired criteria. A GIS used as the basis for a sampling procedure also provides repeatability. Random, stratified random, or cluster sampling can all be accomplished with a GIS.

The powers of GIS extend beyond the data analysis phase as well. Because of the power

Louisiana has a statewide discharger inventory in a GIS. The GIS is a detailed graphical model of the state that contains the location of all known discharges. It is linked to the state Office of Water's databases and EPA's Toxic Release Inventory database by discharger identification number. It includes information on the segment of the water body that is discharged into, and the inventory provides efficient and effective access to a large quantity of data. Since the data can be visually portrayed, the GIS improves comprehension of the impact of waste discharge on the environment, as well as understanding of numerous interrelated waste discharges and their combined impact on large areas (such as entire water quality basins). The GIS also assists in both technical and management decisions (Richards, 1993).

of the data analysis that is possible with a GIS, use of one can lead to improvements in data collection activity design, data tracking methods, database management, and program evaluation. The powerful spatial relationships created through the use of GIS can make data more accessible to a wider audience, thus making GIS a valuable tool for the communication of results of surveys and analyses, and the ability to select from a variety of data elements for data analysis permits customizing the analysis of data for a variety of audiences.

2.4 Summary of Program Elements for a Successful BMP Inventory

The essential elements of a successful urban BMP compliance tracking program include the following:

- Clear and specific program goals
- Technical guidelines for site evaluation, design, construction, and operation
- Regular system monitoring
- Licensing or certification of all service providers
- Effective enforcement mechanisms
- Appropriate incentives
- Adequate records management.

Conversely, the four primary reasons that urban BMP programs fail are insufficient funding; programs that are inappropriate for the specific circumstances under which they are to be implemented; lack of monitoring, inspection, and program evaluation; and lack of public education (USEPA, 1997a). An effective BMP implementation tracking program will generate considerable data and information regarding existing, new, and upgraded BMPs. Essential data management elements include data collection, database development, data entry, possibly data geocoding, and data analysis.

It is not always possible to track the implementation of every BMP of interest. Sampling a subpopulation and extrapolating the findings to the entire population may be preferred due to time, funding, or personnel constraints. Lack of adequate legal authorities might also hinder the collection of data sufficient to track BMP implementation. If an inventory of all BMPs of interest is not possible, care should be taken to prepare a statistically valid sampling plan as discussed in Chapter 3.

CHAPTER 3. SAMPLING DESIGN

3.1 INTRODUCTION

This chapter discusses recommended methods for designing sampling programs to track and evaluate the implementation of nonpoint source control measures. This chapter does not address sampling to determine whether management measures (MMs) or best management practices (BMPs) are effective, since no water quality sampling is done. Because of the variation in urban practices and related nonpoint source control measures implemented throughout the United States, the approaches taken by various states to track and evaluate nonpoint source control measure implementation will differ. Nevertheless, all statistical sampling approaches should be based on sound methods for selecting sampling strategies, computing sample sizes, and evaluating data. EPA recommends that states should consult with a trained statistician to be certain that the approach, design, and assumptions are appropriate to the task at hand.

As described in Chapter 1, implementation monitoring is the focus of this guidance. Effectiveness monitoring is the focus of another guidance prepared by EPA, *Monitoring Guidance for Determining the Effectiveness of Nonpoint Source Controls* (USEPA, 1997). The recommendations and examples in this chapter address two primary monitoring goals:

- Determine the extent to which MMs and BMPs are implemented in accordance with relevant standards and specifications.
- Determine whether there is a change in the extent to which MMs and BMPs are being implemented.

For example, local regulatory personnel might be interested in whether regulations for septic tank inspection and pumping are being adhered to in regions with particular water quality problems. State or county personnel might also be interested in whether, in response to an intensive effort in targeted watersheds to decrease the use of fertilizers and pesticides on residential lawns, there is a detectable change in homeowner behavior.

3.1.1 Study Objectives

To develop a study design, clear, quantitative monitoring objectives must be developed. For example, the objective might be to estimate the percent of local governments that require attenuation of the "first flush" of runoff to within ±5 percent. Or perhaps a state is preparing to perform an extensive 2-year outreach effort to educate citizens on the impacts of improper lawn care. In this case, detecting a 10 percent change in resident's lawn care practices might be of interest. In the first example, summary statistics are developed to describe the current status, whereas in the second example, some sort of statistical analysis (hypothesis testing) is performed to determine whether a significant change has really occurred. This choice has an impact on how the data are collected. As an example, summary statistics might require unbalanced sample allocations to account for variability such as the type of local government, whereas balanced designs (e.g., two sets of data with the same number of observations in each set) are more typical for hypothesis testing.

3.1.2 Probabilistic Sampling

Most study designs that are appropriate for tracking and evaluating implementation are based on a probabilistic approach since tracking every MM or BMP is usually not cost-effective. In a probabilistic approach, individuals are randomly selected from the entire group (see ***Example***). The selected individuals are evaluated, and the results from the individuals provide an unbiased assessment about the entire group. Applying the results from randomly selected individuals to the entire group is *statistical inference*. Statistical inference enables one to determine, for example, in terms of probability, the percentage of local governments that require water quality controls for urban runoff without visiting every community. One could also determine whether a change in homeowners' use of lawn care products is within the range of what could occur by chance or whether it is large enough to indicate a real modification of homeowner behavior.

The group about which inferences are made is the population or *target population*, which consists of *population units*. The *sample*

A survey of the residential population within three small Baltimore, Maryland watersheds was conducted in order to:

- Characterize pesticide usage in the residential areas;
- Test the suitability of sampling locations for future monitoring;
- Obtain stream data to correlate with results of the usage survey; and
- Demonstrate the feasibility of characterizing urban nonpoint source pesticide pollution.

Information for the survey was obtained via door-to-door interviews of randomly selected residents and mail and telephone surveys of commercial pesticide applicators that had been hired by the residents that were interviewed. A total of 484 interviews, or 10 percent of the residential population in the three watersheds, were conducted. The overall response rate to the survey was 69 percent.

The following information was obtained from the survey:

- The percentage of residents that had applied pesticides.
- Where (i.e., indoors and/or outdoors) pesticides were used by the residents.
- The level of use of spray applicators.
- The level of use of fertilizers (i.e., the percentage of residents that used fertilizers, when they were applied, and their frequency of application).
- The use and disposal of petroleum products (i.e., motor oil and antifreeze).
- A listing of brand names of and active ingredients in the pesticides used by the residents.

Example . . . Pesticide usage survey (Kroll and Murphy, 1994).

population is the set of population units that are directly available for measurement. For example, if the objective is to determine the degree to which residents are limiting the use of lawn care products, the population to be sampled would be residential areas with single-family homes or multi-family housing areas with large landscaped areas. Statistical inferences can be made only about the target population available for sampling. For example, if installation of stormwater BMPs is being assessed and only government facilities can be sampled, inferences cannot be made about the management of private lands. Another example to consider is a mail survey. In most cases, only a percentage of survey forms is returned. The extent to which nonrespondents bias the survey findings should be examined: Do the nonrespondents represent those less likely to implement the MM of interest? Typically, a second mailing, phone calls, or visits to those who do not respondent are necessary to evaluate the impact of nonrespondents on the results.

The most common types of sampling that should be used for implementation monitoring are summarized in Table 3-1. In general, probabilistic approaches are preferred. However, there might be circumstances under which targeted sampling should be used. Targeted sampling refers to using best professional judgement for selecting sample locations. For example, state or county regulatory personnel deciding to evaluate all MMs or BMPs in a given watershed would be targeted sampling. The choice of a sampling plan depends on study objectives, patterns of

Table 3-1. Example applications of four sampling designs for implementation monitoring.

Sampling Design	**Example/Applicability**
Simple Random Sampling	Estimate the proportion of homeowners that use herbicides on their lawn. Applicable when there are major patterns in the group of homeowners targeted for the survey.
Stratified Random Sampling	Estimate the proportion of homeowners that use herbicides on their lawn as a function of subdivision. Applicable when herbicide use is expected to be different based on the subdivision or other distinguishing homeowner characteristic (e.g., owner/renter, self/lawn service).
Cluster Sampling	Estimate the proportion of homeowners that use herbicides on their lawn. Applicable when it is more cost effective to sample groups of homeowners rather than individual homeowners. (See Section 3.3.3 for a numerical example comparison to simple random sampling.)
Systematic Sampling	Estimate the proportion of homeowners that use herbicides on their lawn. Applicable when working from a (phone or mailing) list and the list is ordered by some characteristic unrelated to herbicide use.

variability in the target population, cost-effectiveness of alternative plans, types of measurements to be made, and convenience (Gilbert, 1987).

Simple random sampling is the most elementary type of sampling. Each unit of the target population has an equal chance of being selected. This type of sampling is appropriate when there are no major trends, cycles, or patterns in the target population (Cochran, 1977). Random sampling can be applied in a variety of ways, including selection of jurisdictions within a state or BMP sites within a watershed. Random samples can also be taken at different times at a single site. Figure 3-1 provides an example of simple random sampling from a listing of potential inspection sites and from a map.

If the pattern of MM and BMP implementation is expected to be uniform across the study area, simple random sampling is appropriate to estimate the extent of implementation. If, however, implementation is homogeneous only within certain categories (e.g., federal, state, or private lands), stratified random sampling should be used.

In *stratified random sampling*, the target population is divided into groups called strata. Simple random sampling is then used within each stratum. The goal of stratification is to increase the accuracy of the estimated mean values over what could have been obtained using simple random sampling of the entire population. The method makes use of prior information to divide the target population into subgroups that are internally homogeneous. Stratification involves the use of categorical variables to group observations into more units, thereby reducing the variability of observations within each unit. There are a number of ways to "select" sites, or sets of sites (e.g., by type of receiving waterbody, land use, age of BMP, time elapsed since the last inspection or maintenance). For example, in counties with large urban areas and the resources to develop and implement extensive urban runoff management programs, there might be different patterns of BMP implementation than in counties with smaller towns that do not have equivalent resources. Depending on the type of BMPs to be examined (detention ponds versus household waste disposal) different stratification might be necessary. In general, a larger number of samples should be taken in a stratum if the stratum is more variable, larger, or less costly to sample than other strata. For example, if BMP implementation is more variable in less developed areas, a greater number of sampling sites might be needed in that stratum to increase the precision of the overall estimate. Cochran (1977) found that stratified random sampling provides a better estimate of the mean for a population with a trend, followed in order by systematic sampling (discussed later) and simple random sampling. He also noted that stratification typically results in a smaller variance for the estimated mean or total than that which results from comparable simple random sampling.

If the state believes that there will be a difference between two or more subsets of sites, such as between types of development (commercial, residential, etc.), the sites can first be stratified into these subsets and a random sample taken within each subset (McNew, 1990). to be certain that important information will not be lost, or that MM or BMP use will not be misrepresented as a result of treating all potential survey sites as equal.

BMP Catalog No.	Receiving Waterbody	BMP Type	Location Code
1	Stream	OSDS	N3
2	Pond	OSDS	S4
3	Pond	Stormwater	S2
4	Stream	Construction	E5
5	River	Stormwater	S1
6	River	OSDS	S7
7	Lake	Construction	W18
8	Lake	OSDS	E34
• • •	• • •	• • •	• • •
118	Stream	Construction	S21
119	Stream	Construction	W7
120	Pond	Construction	W4
121	River	Stormwater	N5
122	Bay	Construction	N9
123	Bay	OSDS	S3
124	Stream	OSDS	W11
125	Pond	Construction	E14
126	Stream	Construction	S14
127	River	Stormwater	S8
128	Pond	OSDS	N13

Figure 3-1a. Simple random sampling from a listing of BMPs. In this listing, all BMPs are presented as a single list and BMPs are selected randomly from the entire list. Shaded BMPs represent those selected for sampling.

Figure 3-1b. Simple random sampling from a map. Dots represent sites. All sites of interest are represented on the map, and the sites to be sampled (open dots—○) were selected randomly from all of those on the map. The shaded lines on the map could represent county, watershed, hydrologic, or some other boundary, but they are ignored for the purposes of simple random sampling.

It might also be of interest to compare the relative percentages of areas with poor, fair, and good soil percolation that have septic tanks. Areas with poor or fair percolation might be responsible for a larger share of nutrient loadings to ground and surface waters. The region of interest would first be divided into strata based on soil percolation characteristics, and sites within each stratum would be selected randomly to determine the influence of soil type on nutrient enrichment in surface and ground waters. Figure 3-2 provides an example of stratified random sampling from a listing of potential inspection sites and from a map.

Cluster sampling is applied in cases where it is more practical to measure randomly selected groups of individual units than to measure randomly selected individual units (Gilbert, 1987). In cluster sampling, the total population is divided into a number of relatively small subdivisions, or clusters, and then some of the subdivisions are randomly selected for sampling. For one-stage cluster sampling, the selected clusters are sampled totally. In two-stage cluster sampling, random sampling is performed within each cluster (Gaugush, 1987). For example, this approach might be useful if a state wants to estimate the areas within environmentally sensitive watersheds where additional pretreatment of urban runoff might be needed. All areas within the watershed with 30 percent or more of the land zoned for commercial use might be regarded as a single cluster. Once all clusters have been identified, specific clusters can be randomly chosen for sampling. Freund (1973) notes that estimates based on cluster sampling are generally not as good as those based on simple random samples, but they are more cost-effective. Gaugush (1987) believes that the difficulty associated with analyzing cluster samples is compensated for by the reduced sampling cost. Figure 3-3 provides an example of cluster sampling from a listing of potential inspection sites and from a map.

Systematic sampling is used extensively in water quality monitoring programs because it is relatively easy to do from a management perspective. In systematic sampling the first sample has a random starting point and each subsequent sample has a constant distance from the previous sample. For example, if a sample size of 70 is desired from a mailing list of 700 gas station operators, the first sample would be randomly selected from among the first 10 people, say the seventh person. Subsequent samples would then be based on the 17th, 27th, ..., 697th person. In contrast, a stratified random sampling approach for the same case might involve sorting the mailing list by county and then randomly selecting gas station operators from each county. Figure 3-4 provides an example of systematic sampling from a listing of potential inspection sites and from a map.

In general, systematic sampling is superior to stratified random sampling when only one or two samples per stratum are taken for estimating the mean (Cochran, 1977) or when is there is a known pattern of management measure implementation. Gilbert (1987) reports that systematic sampling is equivalent to simple random sampling in estimating the mean if the target population has no trends, strata, or correlations among the population units. Cochran (1977) notes that on the average, simple random sampling and systematic sampling have equal variances. However, Cochran (1977) also states that for any single population for which the number of sampling units is small, the variance from systematic sampling is erratic and might be smaller

BMP Catalog No.	Receiving WaterBody	BMP Type	Location Code
1	Stream	OSDS	N3
2	Pond	OSDS	S4
6	River	OSDS	S7
8	Lake	OSDS	E34
• • •	• • •	• • •	• • •
123	Bay	OSDS	S3
124	Stream	OSDS	W11
128	Pond	OSDS	N13
3	Pond	Stormwater	S2
5	River	Stormwater	S1
• • •	• • •	• • •	• • •
121	River	Stormwater	N5
127	River	Stormwater	S8
4	Stream	Construction	E5
7	Lake	Construction	W18
• • •	• • •	• • •	• • •
118	Stream	Construction	S21
119	Stream	Construction	W7
120	Pond	Construction	W4
122	Bay	Construction	N9
125	Pond	Construction	E14
126	Stream	Construction	S14

Figure 3-2a. Stratified random sampling from a listing of BMPs. Within this listing, BMPs are subdivided by BMP type. Then, considering only one BMP type (e.g., OSDS), some BMPs are selected randomly. The process of random sampling is then repeated for the other BMP types (i.e., stormwater, construction). Shaded BMPs represent those selected for sampling.

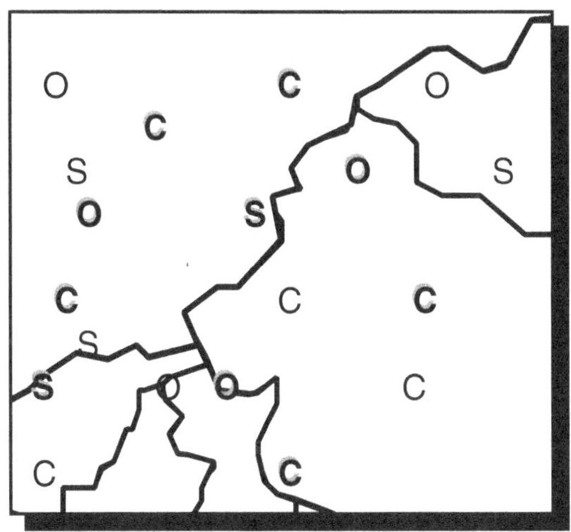

Figure 3-2b. Stratified random sampling from a map. Letters represent sites, subdivided by type (O = OSDS, C = construction, S = stormwater). All sites of interest are represented on the map. From all sites in one type category, some were randomly selected for sampling (**shadowed** sites). The process was repeated for each site type category. The shaded lines on the map could represent counties, soil types, or some other boundary, and could have been used as a means for separating the sites into categories for the sampling process.

BMP Catalog No.	Receiving Waterbody	BMP Type	Location Code/ Residential Zone
1	Stream	OSDS	N3/R1a
121	River	Stormwater	N5/R1a
122	Bay	Construction	N9/R1a
128	Pond	OSDS	N13/R1a
4	Stream	Construction	E5/R1b
8	Lake	OSDS	E34/R1b
125	Pond	Construction	E14/R2a
2	Pond	OSDS	S4/R2a
3	Pond	Stormwater	S2/R2a
5	River	Stormwater	S1/R2a
6	River	OSDS	S7/R2a
118	Stream	Construction	S21/R2b
123	Bay	OSDS	S3/R2b
126	Stream	Construction	S14/R2c
127	River	Stormwater	S8/R3a
7	Lake	Construction	W18/R3a
119	Stream	Construction	W7/R3b
120	Pond	Construction	W4/R3b
124	Stream	OSDS	W11/R3b

Figure 3-3a. One-stage cluster sampling from a listing of BMPs. Within this listing, BMPs are organized by residential zone. Some of the residential zones were then randomly selected and all BMPs in those residential zones were selected for sampling. Shaded BMPs represent those selected for sampling.

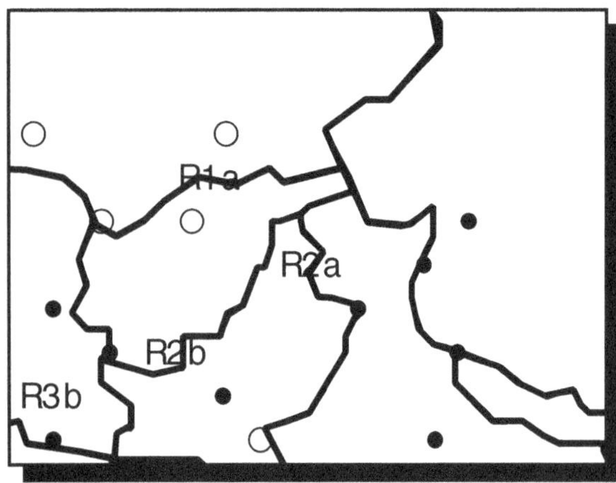

Figure 3-3b. Cluster sampling from a map. All sites in the area of interest are represented on the map (closed {●} and open {○} dots). Residential zones were selected randomly, and all BMPs in those zones (open dots {○}) were selected for sampling. Shaded lines could also have represented another type of boundary, such as soil type, county, or watershed, and could have been used as the basis for the sampling process as well.

BMP Catalog No.	Receiving Waterbody	BMP Type	Location Code
1	Stream	OSDS	N3
2	Pond	OSDS	S4
3	Pond	Stormwater	S2
4	Stream	Construction	E5
5	River	Stormwater	S1
6	River	OSDS	S7
7	Lake	Construction	W18
8	Lake	OSDS	E34
• • •	• • •	• • •	• • •
118	Stream	Construction	S21
119	Stream	Construction	W7
120	Pond	Construction	W4
121	River	Stormwater	N5
122	Bay	Construction	N9
123	Bay	OSDS	S3
124	Stream	OSDS	W11
125	Pond	Construction	E14
126	Stream	Construction	S14
127	River	Stormwater	S8
128	Pond	OSDS	N13

Figure 3-4a. Systematic sampling from a listing of BMPs. From a listing of all BMPs of interest, an initial site (No. 3) was selected randomly from among the first ten on the list. Every fifth BMP listed was subsequently selected for sampling. Shaded BMPs represent those selected for sampling.

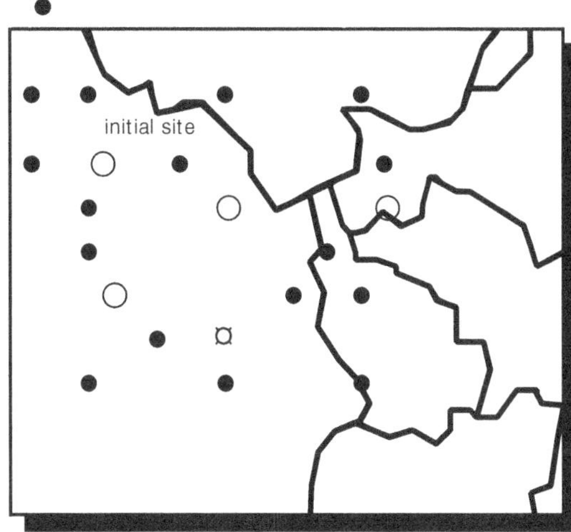

Figure 3-4b. Systematic sampling from a map. Dots (● and ○) represent sites of interest. A single point on the map (¤) and one of the sites were randomly selected. A line was stretched outward from the point to (and beyond) the selected site. The line was then rotated about the map and every fifth dot that it touched was selected for sampling (open dots—○). The direction of rotation was determined prior to selection of the point of the line's origin and the initial site. The shaded lines on the map could represent county boundaries, soil type, watershed, or some other boundary, but were not used for the sampling process.

or larger than the variance from simple random sampling.

Gilbert (1987) cautions that any periodic variation in the target population should be known before establishing a systematic sampling program. Sampling intervals that are equal to or multiples of the target population's cycle of variation might result in biased estimates of the population mean. Systematic sampling can be designed to capitalize on a periodic structure if that structure can be characterized sufficiently (Cochran, 1977). A simple or stratified random sample is recommended, however, in cases where the periodic structure is not well known or if the randomly selected starting point is likely to have an impact on the results (Cochran, 1977).

Gilbert (1987) notes that assumptions about the population are required in estimating population variance from a single systematic sample of a given size. However, there are systematic sampling approaches that do support unbiased estimation of population variance, including multiple systematic sampling, systematic stratified sampling, and two-stage sampling (Gilbert, 1987). In multiple systematic sampling more than one systematic sample is taken from the target population. Systematic stratified sampling involves the collection of two or more systematic samples within each stratum.

3.1.3 Measurement and Sampling Errors

In addition to making sure that samples are representative of the sample population, it is also necessary to consider the types of bias or error that might be introduced into the study. *Measurement error* is the deviation of a measurement from the true value (e.g., the percent of resident participation in "amnesty days" for household hazardous waste was estimated as 60 percent while the true value was 55 percent). A consistent under- or overestimation of the true value is referred to as *measurement bias*. Random *sampling error* arises from the variability from one population unit to the next (Gilbert, 1987), explaining why the proportion of homeowners or developers using a certain BMP differs from one survey to another.

The goal of sampling is to obtain an accurate estimate by reducing the sampling and measurements errors to acceptable levels while explaining as much of the variability as possible to improve the precision of the estimates (Gaugush, 1987). *Precision* is a measure of how close an agreement there is among individual measurements of the same population. The *accuracy* of a measurement refers to how close the measurement is to the true value. If a study has low bias and high precision, the results will have high accuracy. Figure 3-5 illustrates the relationship between bias, precision, and accuracy.

As suggested earlier, numerous sources of variability should be accounted for in developing a sampling design. Sampling errors are introduced by virtue of the natural variability within any given population of interest. As sampling errors relate to MM or BMP implementation, the most effective method for reducing such errors is to carefully determine the target population and to stratify the target population to minimize the nonuniformity in each stratum.

Measurement errors can be minimized by ensuring that interview questions or surveys

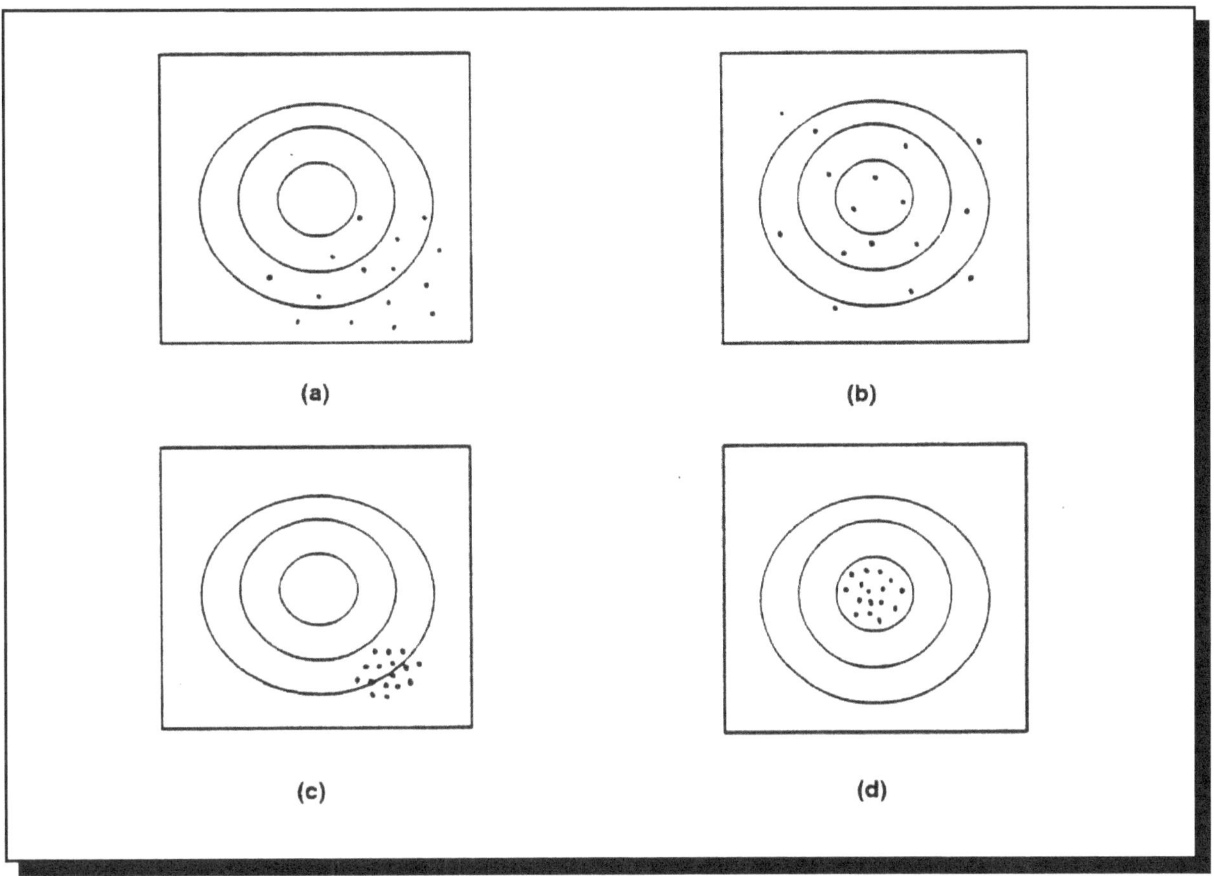

Figure 3-5. Graphical representation of the relationship between bias, precision, and accuracy (after Gilbert, 1987). (a): high bias + low precision = low accuracy; (b): low bias + low precision = low accuracy; (c): high bias + high precision = low accuracy; and (d): low bias + high precision = high accuracy.

are well designed. If a survey is used as a data collection tool, for example, the investigator should evaluate the nonrespondents to determine whether there is a bias in who returned the results (e.g., whether the nonrespondents were more or less likely to implement MMs or BMPs). If data are collected by sending staff out to inspect randomly selected BMPs for operation and maintenance compliance, the approaches for inspecting the BMPs should be consistent. For example, a determination that stormwater ponds are "free of debris" or that swales have been "properly installed" requires consistent interpretation of these terms with respect to actual onsite conditions.

Reducing sampling errors below a certain point (relative to measurement errors) does not necessarily benefit the resulting analysis because total error is a function of the two types of error. For example, if measurement errors such as response or interviewing errors are large, there is no point in taking a huge sample to reduce the sampling error of the estimate since the total error will be primarily determined by the measurement error. Measurement error is of particular concern when homeowner or developer surveys are used for implementation monitoring. Likewise, reducing measurement errors would not be worthwhile if only a small sample size were available for analysis because there would be a large sampling error (and therefore a large total error) regardless of the size of the measurement error. A proper balance between sampling and measurement errors should be maintained because research accuracy limits effective sample size and vice versa (Blalock, 1979).

3.1.4 Estimation and Hypothesis Testing

Rather than presenting every observation collected, the data analyst usually summarizes major characteristics with a few descriptive statistics. Descriptive statistics include any characteristic designed to summarize an important feature of a data set. A *point estimate* is a single number that represents the descriptive statistic. Statistics common to implementation monitoring include proportions, means, medians, totals, and others. When estimating parameters of a population, such as the proportion or mean, it is useful to estimate the *confidence interval.* The confidence interval indicates the range in which the true value lies for a stated confidence level. For example, if it is estimated that 65 percent of structural BMPs are inspected annually and the 90 percent confidence limit is ±5 percent, there is a 90 percent chance that between 60 and 70 percent of BMPs are inspected annually.

Hypothesis testing should be used to determine whether the level of MM and BMP implementation has changed over time. The *null hypothesis* (H_0) is the root of hypothesis testing. Traditionally, H_0 is a statement of no change, no effect, or no difference; for example, "the proportion of developers that implement erosion and sediment control (ESC) BMPs for construction sites after participation in a certification program is equal to the proportion of developers that implement ESC BMPs for construction sites before the certification program." The *alternative hypothesis* (H_a) is counter to H_0, traditionally being a statement of change, effect, or difference. If H_0 is rejected, H_a is accepted. Regardless of the statistical test selected for analyzing the data, the analyst must select the *significance level* (α) of the test. That is, the analyst must determine what error level is acceptable. There are two types of errors in hypothesis testing:

Type I: H_0 is rejected when H_0 is really true.

Type II: H_0 is accepted when H_0 is really false.

Table 3-2 depicts these errors, with the magnitude of Type I errors represented by α and the magnitude of Type II errors represented by β. The probability of making a Type I error is equal to the α of the test and is selected by the data analyst. In most cases, managers or analysts will define $1-\alpha$ to be in the range of 0.90 to 0.99 (e.g., a confidence level of 90 to 99 percent), although there have been applications where $1-\alpha$ has been set to as low as 0.80. Selecting a 95 percent confidence level implies that the analyst will reject the H_0 when H_0 is true (i.e., a false positive) 5 percent of the time. The same notion applies to the confidence interval for point estimates described above: α is set to 0.10, and there is a 10 percent chance that the true percentage of BMPs inspected annually is outside the 60 to

Table 3-2. Errors in hypothesis testing.

Decision	State of Affairs in the Population	
	H_o is True	H_o is False
Accept H_o	1-α (Confidence level)	β (Type II error)
Reject H_o	α (Significance level) (Type I error)	1-β (Power)

70 percent range. This implies that if the decisions to be made based on the analysis are major (i.e., affect many people in adverse or costly ways) the confidence level needs to be greater. For less significant decisions (i.e., low-cost ramifications) the confidence level can be lower.

Type II error depends on the significance level, sample size, and variability, and which alternative hypothesis is true. *Power (1-β)* is defined as the probability of correctly rejecting H_o when H_o is false. In general, for a fixed sample size, α and β vary inversely. For a fixed α, β can be reduced by increasing the sample size (Remington and Schork, 1970).

3.2 Sampling Considerations

In a document of this brevity, it is not possible to address all of the issues that face technical staff who are responsible for developing and implementing studies to track and evaluate the implementation of nonpoint source control measures. For example, the best time to conduct a survey or do onsite visits varies with BMP and type of study. A single time of the year that would be best for all BMPs cannot be identified. Some BMPs can be checked any time of the year, whereas others have a small window of opportunity. In areas that have distinct warm and cold seasons, the warm season might be the most effective time of year to assess the implementation of lawn care BMPs.

The timing of an implementation survey might also depend on actions taken prior to the survey. If the goal of the study is to determine the effectiveness of a public education program, sampling should be timed to ensure that there was sufficient time for outreach activities and for the residents to implement the desired practices. In such a case, telephone calls would be time to reach residents when they are more receptive to participation in a survey, such as during times when they are home but not "busy" (e.g., after dinner).

Another factor that must be considered is that survey personnel must have permission to perform site visits from each affected site owner or developer prior to arriving at the sites. Where access is denied, a replacement site is needed. Replacement sites are selected in accordance with the type of site selection being used, i.e., simple random, stratified random, cluster, or systematic. This can be addressed by requiring site access as part of approval for building codes, permits, etc.

From a study design perspective, all of these issues—study objectives, sampling strategy, allowable error, and formulation of hypotheses—must be considered together when determining the sampling strategy. This section describes common issues that the technical staff might consider in targeting their sampling efforts or determining whether to stratify their sampling efforts. In general, if there is reason to believe that there are different rates of BMP or MM implementation in different groups, stratified random sampling should increase overall accuracy. Following the discussion, a list of resources that can be used to facilitate evaluating these issues is presented.

3.2.1 Urbanized and Urbanizing Areas

The number and type of BMPs currently in use is dependent on, among other things, whether an area is already "built-out" or under development. In areas that are primarily built out (i.e., downtown areas of cities and towns), urban stormwater controls are already in place in some form, although many only address water quantity issues (flood control) and not water quality concerns. There are also space limitations for installing new BMPs. In areas that are undergoing development, urban runoff controls for both quantity and quality can be installed as development occurs. Therefore, sampling can be stratified depending on the level of development in an area. It might be unreasonable to expect that BMPs that require retention of stormwater onsite be implemented in larger cities, however, these areas are very suited to nonstructural controls such as pet waste ordinances, street sweeping, and public education campaigns.

3.2.2 Available Resources and Tax Base

Most structural urban BMPs ultimately fall under the responsibility of the local government. A local government's ability to maintain and operate runoff BMPs depends on a variety of factors, such as staff available for inspection and maintenance, and resources for operation and maintenance. Areas with large populations and/or higher tax bases might be more able to develop and implement an urban runoff control program than urban areas with small populations and low tax bases. Issues to be considered include (1) tax base and percent of tax base dedicated to environmental protection, and (2) size of local government and environmental staff.

3.2.3 Proximity to Sensitive Habitats

The types of urban runoff controls used are often related to the types of resources in need of protection. For example, areas close to sensitive coastal habitats (e.g., shellfish harvesting areas, fish spawning grounds, endangered species habitats) or public water supplies, might require stricter runoff control measures than areas not in the vicinity of such resources.

3.2.4 Federal Requirements

The 1987 amendments to the Clean Water Act included a mandate to regulate storm water point sources. EPA subsequently developed a comprehensive, phased program for controlling urban and industrial storm water discharges. Phase I of the program required areas with municipal separate storm sewer systems (MS4s) serving populations greater than 100,000 to apply for a National Pollutant Discharge Elimination System (NPDES) permit for their MS4s. These municipal permits specify that urban runoff be controlled to the maximum extent practicable through implementation of a variety of measures and include sampling to characterize the discharges from MS4s as well as ongoing monitoring of storm water quality to assess program effectiveness and to ensure compliance. The Phase I NPDES Storm Water program also applies to discharges associated with industrial activity, including construction sites disturbing 5 acres or more.

The Phase II NPDES Storm Water program is currently under regulatory development. A proposed regulation was published in 1998 and the final rule is anticipated in 1999. Based on the proposed rule, this phase of the program will identify smaller MS4s and certain construction sites smaller than 5 acres for control. At this time, however, in all areas that are not subject to Phase I, control of urban runoff is voluntary (except urban coastal areas subject to CZARA). Therefore, smaller, noncoastal urban areas might not be implementing urban runoff BMPs at the same level as larger and coastal urban areas.

3.2.5 Sources of Information

For a truly random selection of population units, it is necessary to access or develop a database that includes the entire target population. U.S. Census data can help identify the population, and therefore level of development, for certain areas.

The following are possible sources of information for site selection. Positive and negative attributes of each information source are included.

County Land Maps. These maps can provide information on landowners and possibly land use. County infrastructure maps might have information on the location of stormwater utilities.

U.S. Census Bureau. Part of the Department of Commerce, the Census Bureau is responsible for compiling data and information on a variety of topics, including population, businesses, employment, trade, and tax base. The data are organized and analyzed in several different ways, such as by state, county, and major metropolitan area. The Census Bureau also performs statistical analyses on the data so that they can be useful for a variety of purposes, such as determining rate of change of population in a specific geographic area. The Census Bureau also provides information on areas that are serviced by central sewage collection and treatment systems and areas that are unsewered. This information can help state and local governments focus efforts for monitoring implementation of the MMs for onsite disposal systems.

Complaint Records. Complaint records could be used in combination with other sources. Such records represent areas that have had problems in the past, which will very likely skew the data set.

Local Government Permits. Local governments usually require permits for new development or redevelopment. The information required to obtain a permit, the level of detail contained in the permits, and the extent to which the permit is monitored varies among local governments. At a minimum, it can be determined whether erosion and sediment controls are part of the site grading plan and stormwater management facilities are included in the overall site development plan. Local governments might require inspection, maintenance, and monitoring as conditions of permit issuance.

Public Health Departments. Local departments of public health might maintain records of onsite OSDS inspections, pumping, and maintenance. These records might contain information on soil tests, system design, maintenance history, permit conditions, and inspection results. In areas where water quality problems due to septic systems are a concern, the systems might be monitored on a watershed basis.

3.3 SAMPLE SIZE CALCULATIONS

This section describes methods for estimating sample sizes to compute point estimates such as proportions and means, as well as detecting changes with a given significance level. Usually, several assumptions regarding data distribution, variability, and cost must be made to determine the sample size. Some assumptions might result in sample size estimates that are too high or too low. Depending on the sampling cost and cost for not sampling enough data, it must be decided whether to make conservative or "best-value" assumptions. Because the cost of visiting any individual site or group of sites is relatively constant, it is more economical to collect a few extra samples during the initial visit rather than to realize later on that it is necessary to return to the site(s) to collect additional data. In most cases, the analyst should probably consider evaluating a range of assumptions on the impact of sample size and overall program cost.

To maintain document brevity, some terms and definitions that will be used in the remainder of this chapter are summarized in Table 3-3. These terms are consistent with those in most introductory-level statistics texts, and more information can be found there. Those with some statistical training will note that some of these definitions include an additional term referred to as the *finite population correction term* $(1-\phi)$, where ϕ is equal to n/N. In many applications, the number of population units in the sample population (N) is large in comparison to the population units sampled (n) and $(1-\phi)$ can be ignored. However, depending on the number of units (towns with populations that fall within a certain range, for example) in a particular population, N can become quite small. N is determined by the definition of the sample population and the corresponding population units. If ϕ is greater than 0.1, the finite population correction factor should not be ignored (Cochran, 1977).

Applying any of the equations described in this section is difficult when no historical data set exists to quantify initial estimates of proportions, standard deviations, means, coefficients of variation, or costs. To estimate these parameters, Cochran (1977) recommends four sources:

- Existing information on the same population or a similar population.

- A two-step sample. Use the first-step sampling results to estimate the needed factors, for best design, of the second step. Use data from both steps to estimate the

Table 3-3. Definitions used in sample size calculation equations.

Symbol		Definition
N	=	total number of population units in sample population
n	=	number of samples
n_0	=	preliminary estimate of sample size
a	=	number of successes
p	=	proportion of successes
q	=	proportion of failures (1-p)
x_i	=	i^{th} observation of a sample
$\bar{x}$	=	sample mean
s^2	=	sample variance
s	=	sample standard deviation
$N\bar{x}$	=	total amount
μ	=	population mean
σ^2	=	population variance
σ	=	population standard deviation
C_v	=	coefficient of variation
$s^2(\bar{x})$	=	variance of sample mean
ϕ	=	n/N (unless otherwise stated in text)
$s(\bar{x})$	=	standard error (of sample mean)
1-ϕ	=	finite population correction factor
d	=	allowable error
d_r	=	relative error
Z_α	=	value corresponding to cumulative area of 1-α using the normal distribution (see Table A1).
$t_{\alpha,df}$	=	value corresponding to cumulative area of 1-α using the student t distribution with df degrees of freedom (see Table A2).

$$p = a/n \qquad q = 1-p$$

$$\bar{x} = \frac{1}{n}\sum_{i=1}^{n} x_i \qquad s^2 = \frac{1}{n-1}\sum_{i=1}^{n}(x_i-\bar{x})^2$$

$$s = \sqrt{s^2} \qquad C_v = s/\bar{x}$$

$$d = |\bar{x}-\mu| \qquad d_r = \frac{|\bar{x}-\mu|}{\mu}$$

$$s^2(\bar{x}) = \frac{s^2}{n}(1-\phi) \qquad s(\bar{x}) = \frac{s}{\sqrt{n}}(1-\phi)^{0.5}$$

$$s(N\bar{x}) = \frac{Ns}{\sqrt{n}}(1-\phi)^{0.5} \qquad s(p) = \sqrt{\frac{pq}{n}}(1-\phi)^{0.5}$$

final precision of the characteristic(s) sampled.

- A "pilot study" on a "convenient" or "meaningful" subsample. Use the results to estimate the needed factors. Here the results of the pilot study generally cannot be used in the calculation of the final precision because often the pilot sample is not representative of the entire population to be sampled.
- Informed judgment, or an educated guess.

It is important to note that this document only addresses estimating sample sizes with traditional parametric procedures. The methods described in this document should be appropriate in most cases, considering the type of data expected. If the data to be sampled are skewed, as—for example—water quality data often are, the investigator should plan to transform the data to something symmetric, if not normal, before computing sample sizes (Helsel and Hirsch, 1995). Kupper and Hafner (1989) also note that some of these equations tend to underestimate the necessary sample because power is not taken into consideration. Again, EPA recommends that if you do not have a background in statistics, you should consult with a trained statistician to be certain that your approach, design, and assumptions are appropriate to the task at hand.

What sample size is necessary to estimate the proportion of local governments that implement pet waste disposal ordinances to within ± 5 percent?

What sample size is necessary to estimate the proportion of local governments that implement pet waste disposal ordinances so that the relative error is less than 5 percent?

Although each agency might have specialized tracking requirements, there might be core questions that are common among a number of agencies. Therefore, it is recommended that local agencies integrate their tracking effort with other agencies so that their results can be compared. Local agencies, initiating a tracking program, at a minimum should contact an appropriate state agency to determine whether the goals and sampling procedures from the state or another local agency can be adopted. Note, that even if two programs have the same goal, sampling differences could still result in the data be incomparable.

3.3.1 Simple Random Sampling

In simple random sampling, it is presumed that the sample population is relatively homogeneous and a difference in sampling costs or variability would not be expected. If the cost or variability of any group within the sample population were different, it might be more appropriate to consider a stratified random sampling approach.
To estimate the proportion of local governments that implement a certain BMP or MM such that the allowable error, *d*, meets the study precision requirements (i.e., the true proportion lies between *p-d* and *p+d* with a $1-\alpha$ confidence level), a preliminary estimate of sample size can be computed as (Snedecor and Cochran, 1980)

$$n_o = \frac{(Z_{1-\alpha/2})^2 pq}{d^2} \qquad (3\text{-}1)$$

If the proportion is expected to be a low number, using a constant allowable error might not be appropriate. Ten percent plus/minus 5 percent has a 50 percent relative error. Alternatively, the relative error, d_r, can be specified (i.e., the true proportion lies between $p-d_r p$ and $p+d_r p$ with a $1-\alpha$ confidence level) and a preliminary estimate of sample size can be computed as (Snedecor and Cochran, 1980)

$$n_o = \frac{(Z_{1-\alpha/2})^2 q}{d_r^2 p} \qquad (3\text{-}2)$$

In both equations, the analyst must make an initial estimate of *p* before starting the study. In the first equation, a conservative sample size can be computed by assuming *p* equal to 0.5. In the second equation the sample size gets larger as *p* approaches zero (0) for constant d_r, thus an informed initial estimate of *p* is needed. Values of α typically range from

Case Study: Delaware Method

The Delaware Department of Natural Resources and Environmental Control (DNREC) has developed a methodology (referred to as the *Delaware Method*) to evaluate erosion and sedimentation control program implementation and effectiveness (Piorko, et al., In press). The method involves a numerical ranking of construction site conditions, soil erosion practices and land covers, and sediment control practices, to assess the effectiveness of site controls.

The first step in the Delaware Method involves using a statistical sampling approach to select a valid representative sample from the total population. For example, in New Castle County, Delaware, 40 sites were randomly chosen from a total population of 453 active construction sites. The sample size used by DNREC is consistent with the sample size estimated using equations 3-1 and 3-3 of this guidance. This sample, selected based on the methods from Walpole and Meyers (1972, in Maxted, 1996), yielded a probability estimate of 0.25±0.11 (90 percent confidence interval) of construction sites implementing ESC BMPs in accordance with county permitting requirements. The sample compared favorably with the total population: of the 40 sites randomly selected, 47.5 percent were non-residential, compared with 48.6 percent (220 sites) of the 453 active construction sites that were non-residential. Using the method, DNREC can estimate countywide implementation of selected management measures.

Once the representative sample is developed, the Delaware Method uses a series of worksheets to rank score the sample of construction sites. These worksheets, which are completed in the field, are designed to specifically evaluate the erosion and sedimentation control practices used at a site. The information recorded on these worksheets is used to construct a chart that tabulates the final numerical rating for the total site area and the estimated total tons of soil lost on that site. The results from the numerical ranking method allow for comparison of the effectiveness of erosion and sedimentation control practices among construction sites.

0.01 to 0.10. The final sample size is then estimated as (Snedecor and Cochran, 1980)

$$n = \begin{cases} \dfrac{n_0}{1+\phi} & for\ \phi > 0.1 \\ n_o & otherwise \end{cases} \qquad (3\text{-}3)$$

where ϕ is equal to n_o/N. Table 3-4 demonstrates the impact on n of selecting p, α, d, d_r, and N. For example, 151 random samples are needed to estimate the proportion of 500 households that dispose of household hazardous waste safely to within ±5 percent (d=0.05) with a 95 percent confidence level assuming roughly one-half of households dispose of their hazardous waste safely. Suppose the goal is to estimate the average storage volume of extended detention ponds. (This goal might only be appropriate in areas

> *What sample size is necessary to estimate the average storage volume of extended detention ponds to within ± 1,000 ft^3 per acre of impervious area?*
>
> *What sample size is necessary to estimate the average storage volume of extended detention ponds to within ±10 percent?*

Table 3-4. Comparison of sample size as a function of p, α, d, d_r, and N for estimating proportions using equations 3-1 through 3-3.

Probability of Success, p	Significance level, α	Allowable error, d	Relative error, d_r	Preliminary sample size, n_o	Sample Size, n — Number of Population Units in Sample Population, N				
					500	750	1,000	2,000	Large N
0.1	0.05	0.050	0.500	138	108	117	121	138	138
0.1	0.05	0.075	0.750	61	55	61	61	61	61
0.5	0.05	0.050	0.100	384	217	254	278	322	384
0.5	0.05	0.075	0.150	171	127	139	146	171	171
0.1	0.10	0.050	0.500	97	82	86	97	97	97
0.1	0.10	0.075	0.750	43	43	43	43	43	43
0.5	0.10	0.050	0.100	271	176	199	213	238	271
0.5	0.10	0.075	0.150	120	97	104	107	120	120

that do not have regulations mandating pond size.) The number of random samples required to achieve a desired margin of error when estimating the mean (i.e., the true mean lies between $\bar{x}$-d and $\bar{x}$+d with a 1-α

$$n = \frac{(t_{1-\alpha/2,n-1}s/d)^2}{1 + (t_{1-\alpha/2,n-1}s/d)^2/N} \tag{3-4}$$

confidence level) is (Gilbert, 1987)
If N is large, the above equation can be simplified to

$$n = (t_{1-\alpha/2,n-1}s/d)^2 \tag{3-5}$$

Since the Student's t value is a function of n, Equations 3-4 and 3-5 are applied iteratively. That is, guess at what n will be, look up $t_{1-\alpha/2,n-1}$ from Table A2, and compute a revised n. If the initial guess of n and the revised n are different, use the revised n as the new guess, and repeat the process until the computed value of n converges with the guessed value. If the population standard deviation is known (not too likely), rather than estimated, the above equation can be further simplified to:

$$n = (Z_{1-\alpha/2}\sigma/d)^2 \tag{3-6}$$

To keep the relative error of the mean estimate below a certain level (i.e., the true mean lies between $\bar{x}$-$d_r\bar{x}$ and $\bar{x}$+$d_r\bar{x}$ with a 1-α confidence level), the sample size can be computed with (Gilbert, 1987)
C_v is usually less variable from study to study than are estimates of the standard deviation,

$$n = \frac{(t_{1-\alpha/2,n-1}C_v/d_r)^2}{1+(t_{1-\alpha/2,n-1}C_v/d_r)^2/N} \tag{3-7}$$

which are used in Equations 3-4 through 3-6. Professional judgment and experience, typically based on previous studies, are required to estimate C_v. Had C_v been known, $Z_{1-\alpha/2}$ would have been used in place of $t_{1-\alpha/2,n-1}$ in Equation 3-7. If N is large, Equation 3-7 simplifies to:

$$n = (t_{1-\alpha/2,n-1} C_v / d_r)^2 \qquad (3\text{-}8)$$

Consider a state, for example, where subdivision developments for single-family homes typically range in size from 100 to 1,500 lots, although most have fewer than 400. The goal of the sampling program is to estimate the average storage volume of extended detention ponds. However, the investigator is concerned about skewing the mean estimate with the few large developments. As a result, the sample population for this analysis is the 250 developments with fewer than 400 lots. The investigator also wants to keep the relative error under 15 percent (i.e., $d_r < 0.15$) with a 90 percent confidence level.

Unfortunately, this is the first study of this type that has been done in this state and there is no information about the coefficient of variation, C_v. The investigator, however, has done several site inspections over the last 5 years. Based on this experience, the investigator knows that developers typically build ponds that range in size from 5,000 to 20,000 ft^3. Using this information, the investigator roughly estimates s as (20,000-5,000)/2 or 7,500 (Sanders et al., 1983) and $\bar{x}$ as 12,500. C_v is then estimated as 7,500/12,500, or 0.6. As a first-cut approximation, Equation 3-6 is applied with $Z_{1-\alpha/2}$ equal to 1.645 and assuming N is large:

$$n = (1.645 \times 0.6/0.15)^2$$
$$= 43.3 \approx 44 \textit{ samples}$$

Since n/N is greater than 0.1 and C_v is estimated (i.e., not known), it is best to reestimate n with Equation 3-7 using 44 samples as the initial guess of n. In this case, $t_{1-\alpha/2,n-1}$ is obtained from Table A2 as 1.6811.

$$n = \frac{(1.6811 \times 0.6/0.15)^2}{1 + (1.6811 \times 0.6/0.15)^2/250}$$
$$= 38.3 \approx 39 \textit{ samples}$$

Notice that the revised sample is somewhat smaller than the initial guess of n. In this case it is recommended to reapply Equation 3-7 using 39 samples as the revised guess of n. In this case, $t_{1-\alpha/2,n-1}$ is obtained from Table A2 as 1.6850.

$$n = \frac{(1.6850 \times 0.6/0.15)^2}{1 + (1.6850 \times 0.6/0.15)^2/250}$$
$$= 38.5 \approx 39 \textit{ samples}$$

Since the revised sample size matches the estimated sample size on which $t_{1-\alpha/2,n-1}$ was based, no further iterations are necessary. The proposed study should include 39 developments randomly selected from the 250 developments with fewer than 400 lots.

When interest is focused on whether the level of BMP implementation has changed, it is necessary to estimate the extent of implementation at two different time periods. Alternatively, the proportion from two different populations can be compared. In either case, two independent random samples are taken and a hypothesis test is used to determine whether there has been a signif-icant change in implementation. (See Snedecor and Cochran (1980) for sample size calculations for matched data.) Consider an example in which the proportion of house-holds that properly dispose of household hazardous waste will be estimated at two time periods. What sample size is needed?

What sample size is necessary to determine whether there is a 20 percent difference in household hazardous waste disposal before and after an education program?

What sample size is necessary to detect a difference of 2,000 ft³ per acre of impervious area in average pond storage volume between land owners that plan and develop their own land versus those that hire independent consultants?

To compute sample sizes for comparing two proportions, p_1 and p_2, it is necessary to provide a best estimate for p_1 and p_2, as well as specifying the significance level and power (*1-β*). Recall that power is equal to the probability of rejecting H_0 when H_0 is false. Given this information, the analyst substitutes these values into (Snedecor and Cochran, 1980)

$$n_o = (Z_\alpha + Z_{2\beta})^2 \frac{(p_1 q_1 + p_2 q_2)}{(p_2 - p_1)^2} \qquad (3\text{-}9)$$

where Z_α and $Z_{2\beta}$ correspond to the normal deviate. Although this equation assumes that N large, it is acceptable for practical use (Snedecor and Cochran, 1980). Common values of $(Z_\alpha \text{ and } Z_{2\beta})^2$ are summarized in Table 3-5. To account for p_1 and p_2 being estimated, Z could be substituted with t. In lieu of an iterative calculation, Snedecor and Cochran (1980) propose the following approach: (1) compute n_o using Equation 3-9; (2) round n_o up to the next highest integer, f; and (3) multiply n_o by $(f+3)/(f+1)$ to derive the final estimate of n.

To detect a difference in proportions of 0.20 with a two-sided test, α equal to 0.05, $1-\beta$ equal to 0.90, and an estimate of p_1 and p_2 equal to 0.4 and 0.6, n_o is computed as

$$n_o = 10.51 \frac{[(0.4)(0.6) + (0.6)(0.4)]}{(0.6 - 0.4)^2}$$
$$= 126.1$$

Table 3-5. Common values of $(Z_\alpha + Z_{2\beta})^2$ for estimating sample size for use with equations 3-9 and 3-10.

Power, 1-β	α for One-sided Test			α for Two-sided Test		
	0.01	**0.05**	**0.10**	**0.01**	**0.05**	**0.10**
0.80	10.04	6.18	4.51	11.68	7.85	6.18
0.85	11.31	7.19	5.37	13.05	8.98	7.19
0.90	13.02	8.56	6.57	14.88	10.51	8.56
0.95	15.77	10.82	8.56	17.81	12.99	10.82
0.99	21.65	15.77	13.02	24.03	18.37	15.77

Rounding 126.1 to the next highest integer, *f* is equal to 127, and *n* is computed as 126.1 x 130/128 or 128.1. Therefore 129 samples in each random sample, or 258 total samples, are needed to detect a difference in proportions of 0.2. Beware of other sources of information that give significantly lower estimates of sample size. In some cases the other sources do not specify $1\text{-}\beta$; otherwise, be sure that an "apples-to-apples" comparison is being made.

To compare the average from two random samples to detect a change of δ (i.e., $\bar{x}_2\text{-}\bar{x}_1$), the following equation is used:

$$n_o = (Z_\alpha + Z_{2\beta})^2 \frac{(s_1^2 + s_2^2)}{\delta^2} \qquad (3\text{-}10)$$

Common values of $(Z_\alpha \text{ and } Z_{2\beta})^2$ are summarized in Table 3-5. To account for s_1 and s_2 being estimated, *Z* should be replaced with *t*. In lieu of an iterative calculation, Snedecor and Cochran (1980) propose the following approach: (1) compute n_o using Equation 3-10; (2) round n_o up to the next highest integer, *f*; and (3) multiply n_o by $(f+3)/(f+1)$ to derive the final estimate of *n*.

Continuing the extended detention pond example, where *s* was estimated as 7,500 ft^3, the investigator will also want to compare the average pond size between land owners that plan and develop their own land versus those that hire independent consultants. The investigator believes that it will be necessary to detect a 4,000 ft^3 difference to make an impact on planning decisions. Although the standard deviation might differ between the two groups, there is no particular reason to propose a different *s* at this point. To detect a difference of 4,000 ft^3 with a two-sided test, α equal to 0.05, $1\text{-}\beta$ equal to 0.90, and an estimate of s_1 and s_2 equal to 7,500, n_o is computed as

$$n_o = 10.51 \frac{(7{,}500^2 + 7{,}500^2)}{4000^2} = 73.9 \qquad (3\text{-}11)$$

Rounding 73.9 to the next highest integer, *f* is equal to 74, and *n* is computed as 73.9 x 77/75 or 75.9. Therefore, 76 samples in each random sample, or 152 total samples, are needed to detect a difference of 4,000 ft^3.

3.3.2 Stratified Random Sampling

The key reason for selecting a stratified random sampling strategy over simple random sampling is to divide a heterogeneous population into more homogeneous groups. If populations are grouped based on size (e.g., lawn size) when there is a large number of small units and a few larger units, a large gain in precision can be expected (Snedecor and Cochran, 1980). Stratifying also allows the investigator to efficiently allocate sampling resources based on cost. Information from preliminary studies (see Section 3.3) can provide useful sampling cost information. The stratum mean, $\bar{x}_h$, is computed using the standard approach for estimating the mean.

The overall mean, $\bar{x}_{st}$, is computed as

$$\bar{x}_{st} = \sum_{h=1}^{L} W_h \bar{x}_h \qquad (3\text{-}12)$$

What sample size is necessary to estimate the average number of households that carefully monitor their fertilizer applications when there is a wide variety of lawn sizes?

where L is the number of strata and W_h is the relative size of the hth stratum. W_h can be computed as N_h/N where N_h and N are the number of population units in the hth stratum and the total number of population units across all strata, respectively. Assuming that simple random sampling is used within each stratum, the variance of $\overline{x}_{st}$ is estimated as (Gilbert, 1987)

$$s^2(\overline{x_{st}}) = \frac{1}{N^2}\sum_{h=1}^{L} N_h^2\left(1-\frac{n_h}{N_h}\right)\frac{s_h^2}{n_h} \quad (3\text{-}13)$$

where n_h is the number of samples in the hth stratum and s_h^2 is computed as (Gilbert, 1987)

$$s_h^2 = \frac{1}{n_h-1}\sum_{i=1}^{n_h}(x_{h,i}-\overline{x}_h)^2 \quad (3\text{-}14)$$

There are several procedures for computing sample sizes. The method described below allocates samples based on stratum size, variability, and unit sampling cost. If $s^2(\overline{x}_{st})$ is specified as V for a design goal, n can be obtained from (Gilbert, 1987)

$$n = \frac{\left(\sum_{h=1}^{L} W_h s_h\sqrt{c_h}\right)\sum_{h=1}^{L} W_h s_h/\sqrt{c_h}}{V+\frac{1}{N}\sum_{h=1}^{L} W_h s_h^2} \quad (3\text{-}15)$$

where c_h is the per unit sampling cost in the hth stratum and n_h is estimated as (Gilbert, 1987)

$$n_h = n\,\frac{W_h s_h/\sqrt{c_h}}{\sum_{h=1}^{L} W_h s_h/\sqrt{c_h}} \quad (3\text{-}16)$$

In the discussion above, the goal is to estimate an overall mean. To apply a stratified random sampling approach to estimating proportions, substitute p_h, p_{st}, p_hq_h, and $s^2(p_{st})$ for $\overline{x}_h$, $\overline{x}_{st}$, s_h^2, and $s^2(\overline{x}_{st})$ in the above equations, respectively.

To demonstrate the above approach, consider a local government that wishes to determine the percentage of homeowners in single family residences that implement recommended lawn care practices. The investigator anticipated that there might be a difference in implementation between home owners that do their own work versus households that use lawn care services. Based on some preliminary work, she determined that homeowner perform their own lawn care for 7,000 households while lawn services perform the work for 3,000 households. Table 3-6 presents three basic scenarios for estimating sample size. In the first scenario, sh and ch are assumed equal among all strata. That is, the variability in each of the two groups is expected to be the same, and the cost to complete the survey for one household is the same regardless of group. Using a design goal of V equal to 0.0025 and applying Equation 3-15 yields a total sample size of 99. Since sh and ch are equal, these samples are allocated proportionally to Wh, which is referred to as proportional allocation. This allocation can be verified by comparing the percent sample allocation to Wh. Due to rounding up, a total of 100 samples are allocated.

Under the second scenario, referred to as the Neyman allocation, the variability between strata changes, but unit sample cost is constant. In this example, sh decreases from 0.40 to 0.75 between strata. (This difference is for illustrative purposes and might not be realized

Table 3-6. Allocation of samples.

Who Provides Lawn Care	Number of Lots (N_h)	Relative Size (W_h)	Standard Deviation (s_h)	Unit Sample Cost (c_h)	Sample Allocation	
					Number	%
A) Proportional allocation (s_h and c_h are constant)						
Homeowner	7,000	0.70	0.50	1	70	70.0
Lawn Service	3,000	0.30	0.50	1	30	30.0
Using Equation 3-15, n is equal to 99.0. Applying Equation 3-16 to each stratum yields a total of 100 samples after rounding up to the next integer.						
B) Neyman allocation (c_h is constant)						
Homeowner	7,000	0.70	0.40	1	56	55.4
Lawn Service	3,000	0.30	0.75	1	45	44.6
Using Equation 3-15, n is equal to 100.9. Applying Equation 3-16 to each stratum yields a total of 101 samples after rounding up to the next integer.						
C) Allocation where s_h and c_h are not constant						
Homeowner	7,000	0.70	0.40	1	62	60.2
Lawn Service	3,000	0.30	0.75	1.5	41	39.8
Using Equation 3-15, n is equal to 101.9. Applying Equation 3-16 to each stratum yields a total of 103 samples after rounding up to the next integer.						

in practice.) The total number of samples remained roughly the same; however, an increased number of samples are required for lawn care services. Using proportional allocation 30 percent of the samples are taken from households that use lawn care services whereas approximately 44.6 percent of the samples are taken in the same stratum using the Neyman allocation.

Finally, introducing sample cost variation will also affect sample allocation. In the last scenario it was assumed that it is 50 percent more expensive to evaluate a lot from the stratum that corresponds to the households that use lawn care services. (This difference is for illustrative purposes and might not be realized in practice.) In this example, roughly the same total number of samples are needed to meet the design goal, yet fewer samples are now required from households that use lawn services.

3.3.3 Cluster Sampling

Cluster sampling is commonly used when there is a choice between the size of the sampling unit (e.g., subdivision versus individual residences). In general, it is cheaper to sample larger units than smaller units, but these results tend to be less accurate (Snedecor and Cochran, 1980). Thus, if there is not a unit sampling cost advantage to cluster sampling, it is probably better to use simple random sampling. To decide whether to perform a cluster sample, it will probably be necessary to perform a special investigation to quantify sampling errors and costs using the two approaches.

Perhaps the best approach to explaining the difference between simple random sampling and cluster sampling is to consider an example set of results. In this example, the investigator did an evaluation of BMP implementation to evaluate whether certain practices had been implemented. Since the county was quite large and random sampling costs would be high due to travel time, the investigator stopped at 30 sites (locations). At each site he inspected 10 neighboring residences. In addition to determining whether recommended lawn care practices were being implemented, the investigator would probably collect ancillary data such as whether the household used a lawn care service. For the purposes of explaining cluster sampling, the type of lawn care provider is not critical although it might be in practice. Table 3-7 presents the number of residences (out of 10) at each site that were implementing recommended lawn care practices. At Site 1, for example, 3 of the 10 households were implementing recommended lawn care practices. For the 30 sites, the overall mean is 5.6; a little more than one-half of the residences have implemented recommended lawn care practices. Note that since the population unit corresponds to the 10 residences at each site collectively, thus there are 30 samples and the standard error for the proportion of residences using recommended BMPs is 0.035. Had the investigator incorrectly calculated the standard error using the random sampling equations, he would have computed 0.0287, nearly a 20 percent error.

Since the standard error from the cluster sampling example is 0.035, it is possible to estimate the corresponding simple random sample size to get the same precision using Is collecting 300 samples using a cluster sampling approach cheaper than collecting

$$n = \frac{pq}{s(p)^2} = \frac{(0.56)(0.44)}{0.035^2} = 201 \qquad (3\text{-}17)$$

about 200 simple random samples? If so, cluster sampling should be used; otherwise simple random sampling should be used.

3.3.4 Systematic Sampling

It might be necessary to obtain a baseline estimate of the proportion of residences where a certain BMP (e.g., reduced lawn fertilization) is implemented using a mailed questionnaire or phone survey. Assuming a record of homeowners in the city is available in a sequence unrelated to the manner in which the BMP would be implemented (e.g., in alphabetical order by the homeowner's name), a systematic sample can be obtained in the following manner (Casley and Lury, 1982):

Table 3-7. Number of residences at each site implementing recommended lawn care practices (10 residences inspected at each site).

Site 1: 3[a]	Site 2: 9	Site 3: 5	Site 4: 7	Site 5: 6	Site 6: 4
Site 7: 6	Site 8: 3	Site 9: 5	Site 10: 5	Site 11: 5	Site 12: 7
Site 13: 7	Site 14: 4	Site 15: 7	Site 16: 5	Site 17: 3	Site 18: 8
Site 19: 4	Site 20: 6	Site 21: 8	Site 22: 4	Site 23: 7	Site 24: 4
Site 25: 5	Site 26: 3	Site 27: 3	Site 28: 9	Site 29: 9	Site 30: 7

Grand Total = 168 (i.e., 168 of 300 residents used recommended lawn care practices)

$\bar{x}$ **= 5.6** (i.e., 5.6 out of 10 residents used recommended lawn care practices) — $p = 5.6/10 = 0.560$

***s* = 1.923** — $s' = 1.923/10 = 0.1923$

Standard error using cluster sampling: $s(p) = 0.1923/(30)^{0.5} = 0.035$

Standard error if simple random sampling assumption had been incorrectly used:

$s(p) = ((0.56)(1-0.56)/300)^{0.5} = 0.0287$

[a] At Site 1, for example, 3 of the 10 households were implementing recommended lawn care practices.

1. Select a random number *r* between 1 and *n*, where *n* is the number required in the sample.

2. The sampling units are then r, $r + (N/n)$, $r + (2N/n)$, ..., $r + (n-1)(N/n)$, where N is total number of available records.

If the population units are in random order (e.g., no trends, no natural strata, uncorrelated), systematic sampling is, on average, equivalent to simple random sampling.

Once the sampling units (in this case, specific residences) have been selected, questionnaires can be mailed to homeowners or telephone inquiries made about lawn care practices being followed by the homeowners.

3.3.5 Concluding Remarks

In the previous examples the type of questions asked where generally similar yet dramatically different sample sizes were developed. This probably leaves the reader wondering which one to choose. Clearly simple random sampling is the easiest, but might very well leave the investigator with numerous unanswered questions. The primary basis for selecting a design approach should be based on a careful review of study objectives and the discussion in Section 3.1.2 and Table 3-1. As shown in Section 3.3.3, cluster sampling can be a good alternative to simple random sampling when you can demonstrate a sampling cost savings. In both cases, there is no stratification or optimization based on your a priori knowledge about patterns or sampling costs in the target population. When there are critical factors that you are examining or you know that there are group differences among the target population, stratified sampling (Section 3.3.2) should be used to optimize (in a less-variance sense) the precision of population totals. On the other hand if your objective is to compare implementation among two groups, then sample size calculations derived from the Equations 3-9 or 3-10 (Section 3.3.1) should be used.

CHAPTER 4. METHODS FOR EVALUATING DATA

4.1 INTRODUCTION

Once data have been collected, it is necessary to statistically summarize and analyze the data. EPA recommends that the data analysis methods be selected before collecting the first sample. Many statistical methods have been computerized in easy-to-use software that is available for use on personal computers. Inclusion or exclusion in this section does not imply an endorsement or lack thereof by the EPA. Commercial off-the-shelf software that covers a wide range of statistical and graphical support includes SAS, Statistica, Statgraphics, Systat, Data Desk (Macintosh only), BMDP, and JMP. Numerous spreadsheets, database management packages, and other graphics software can also be used to perform many of the needed analyses. In addition, the following programs, written specifically for environmental analyses, are also available:

- SCOUT: A Data Analysis Program, EPA, NTIS Order Number PB93-505303.

- WQHYDRO (WATER QUALITY/HYDROLOGY GRAPHICS/ANALYSIS SYSTEM), Eric R. Aroner, Environmental Engineer, P.O. Box 18149, Portland, OR 97218.

- WQSTAT, Jim C. Loftis, Department of Chemical and Bioresource Engineering, Colorado State University, Fort Collins, CO 80524.

Computing the proportion of construction sites implementing a certain BMP or the average storage volume per acre of impervious area of extended detention ponds follows directly from the equations presented in Section 3.3 and the equations are not repeated here. The remainder of this section is focused on evaluating changes in BMP implementation. The methods provided in this section provide only a cursory overview of the type of analyses that might be of interest. For a more thorough discussion on these methods, the reader is referred to Gilbert (1987), Snedecor and Cochran (1980), and Helsel and Hirsch (1995). Typically, the data collected for evaluating changes will typically come as two or more sets of random samples. In this case, the analyst will test for a shift or step change.

Depending on the objective, it is appropriate to select a one- or two-sided test. For example, if the analyst knows that BMP implementation will only go up as a result of a regulatory or educational program, a one-sided test could be formulated. Alternatively, if the analyst does not know whether implementation will go up or down, a two-sided test is necessary. To simply compare two random samples to decide if they are significantly different, a two-sided test is used. Typical null hypotheses (H_o) and alternative hypotheses (H_a) for one- and two-sided tests are provided below:

<u>One-sided test</u>

H_o: BMP Implementation (Post regulation) ≤ BMP Implementation (Pre regulation)

H_a: BMP Implementation (Post regulation) > BMP Implementation (Pre regulation)

<u>Two-sided test</u>

H_o: BMP Implementation (Post education program) = BMP Implementation (Pre education program)

H_a: BMP Implementation (Post education program) ≠ BMP Implementation (Pre education program)

Selecting a one-sided test instead of a two-sided test results in an increased power for the same significance level (Winer, 1971). That is, if the conditions are appropriate, a corresponding one-sided test is more desirable than a two-sided test given the same α and sample size. The manager and analyst should take great care in selecting one- or two-sided tests.

4.2 Comparing the Means from Two Independent Random Samples

The Student's *t* test for two samples and the Mann-Whitney test are the most appropriate tests for these types of data. Assuming the data meet the assumptions of the *t* test, the two-sample *t* statistic with n_1+n_2-2 degrees of freedom is (Remington and Schork, 1970)

$$t = \frac{(\bar{x}_1 - \bar{x}_2) - \Delta_0}{s_p \sqrt{\frac{1}{n_1} + \frac{1}{n_2}}} \qquad (4\text{-}1)$$

where n_1 and n_2 are the sample size of the first and second data set and $\bar{x}_1$ and $\bar{x}_2$ are the

Tests for Two Independent Random Samples

Test*	Key Assumptions
Two-sample *t*	• Both data sets must be normally distributed • Data sets should have equal variances†
Mann-Whitney	• None

* The standard forms of these tests require independent random samples.
† The variance homogeneity assumption can be relaxed.

estimated means from the first and second data set, respectively. The pooled standard deviation, s_p, is defined by

$$s_p = \left[\frac{s_1^2(n_1 - 1) + s_2^2(n_2 - 1)}{n_1 + n_2 - 2}\right]^{0.5} \qquad (4\text{-}2)$$

where s_1^2 and s_2^2 correspond to the estimated variances of the first and second data set, respectively. The difference quantity (Δ_o) can be any value, but here it is set to zero. Δ_o can be set to a non-zero value to test whether the difference between the two data sets is greater than a selected value. If the variances are not equal, refer to Snedecor and Cochran (1980) for methods for computing the *t* statistic. In a two-sided test, the value from Equation 4-1 is compared to the *t* value from Table A2 with $\alpha/2$ and n_1+n_2-2 degrees of freedom.

The Mann-Whitney test can also be used to compare two independent random samples. This test is very flexible since there are no assumptions about the distribution of either sample or whether the distributions have to be the same (Helsel and Hirsch, 1995). Wilcoxon (1945) first introduced this test for equal-sized samples. Mann and Whitney (1947) modified the original Wilcoxon's test to apply it to different sample sizes. Here, it is determined whether one data set tends to have larger observations than the other.

If the distributions of the two samples are similar except for location (i.e., similar spread and skew), H_a can be refined to imply that the median concentration from one sample is "greater than," "less than," or "not equal to" the median concentration from the second sample. To achieve this greater detail in H_a, transformations such as logs can be used.

Tables of Mann-Whitney test statistics (e.g., Conover, 1980) can be consulted to determine whether to reject H_o for small sample sizes. If n_1 and n_2 are greater than or equal to 10 observations, the test statistic can be computed from the following equation (Conover, 1980):

$$T_1 = \frac{T - n_1 \frac{n+1}{2}}{\sqrt{\frac{n_1 n_2}{n(n-1)} \sum_{i=1}^{n} R_i^2 - \frac{n_1 n_2 (n+1)^2}{4(n-1)}}} \qquad (4\text{-}3)$$

where

- n_1 = number of observations in sample with fewer observations,
- n_2 = number of observations in sample with more observations,
- n = $n_1 + n_2$,
- T = sum of ranks for sample with fewer observations, and
- R_i = rank for the ith ordered observation used in both samples.

T_1 is normally distributed and Table A1 can be used to determine the appropriate quantile. Helsel and Hirsch (1995) and USEPA (1997) provide detailed examples for both of these tests.

4.3 Comparing the Proportions from Two Independent Samples

Consider the example in which the proportion of site inspection violations has been estimated during two time periods to be p_1 and p_2 using sample sizes of n_1 and n_2, respectively. Assuming a normal approximation is valid, the test statistic under a null hypothesis of equivalent proportions (no change) is

$$\frac{p_1 - p_2}{\sqrt{p(1-p)\left(\frac{1}{n_1} + \frac{1}{n_2}\right)}} \qquad (4\text{-}4)$$

where p is a pooled estimate of proportion and is equal to $(x_1+x_2)/(n_1+n_2)$ and x_1 and x_2 are the number of successes during the two time periods. An estimator for the difference in proportions is simply $p_1 - p_2$.

In an earlier example, it was determined that 129 observations in each sample were needed to detect a difference in proportions of 0.20 with a two-sided test, α equal to 0.05, and $1-\beta$ equal to 0.90. Assuming that 130 samples were taken and p_1 and p_2 were estimated from the data as 0.6 and 0.4, the test statistic would be estimated as

$$\frac{0.6 - 0.4}{\sqrt{0.5(0.5)\left(\frac{1}{130} + \frac{1}{130}\right)}} = 3.22 \qquad (4\text{-}5)$$

Comparing this value to the t value from Table A2 ($\alpha/2 = 0.025$, df=258) of 1.96, H_o is rejected.

4.4 Comparing More Than Two Independent Random Samples

The analysis of variance (ANOVA) and Kruskal-Wallis are extensions of the two-sample *t* and Mann-Whitney tests, respectively, and can be used for analyzing more than two independent random samples when the data are continuous (e.g., mean acreage). Unlike the *t* test described earlier, the ANOVA can have more than one factor or explanatory variable. The Kruskal-Wallis test accommodates only one factor, whereas the Friedman test can be used for two factors. In addition to applying one of the above tests to determine if one of the samples is significantly different from the others, it is also necessary to do postevaluations to determine which of the samples is different. This section recommends Tukey's method to analyze the raw or rank-transformed data only if one of the previous tests (ANOVA, rank-transformed ANOVA, Kruskal-Wallis, Friedman) indicates a significant difference between groups. Tukey's method can be used for equal or unequal sample sizes (Helsel and Hirsch, 1995). The reader is cautioned, when performing an ANOVA using standard software, to be sure that the ANOVA test used matches the data. See USEPA (1997) for a more detailed discussion on comparing more than two independent random samples.

4.5 Comparing Categorical Data

In comparing categorical data, it is important to distinguish between nominal categories (e.g., land ownership, county location, type of BMP) and ordinal categories (e.g., BMP implementation rankings, low-medium-high scales).

The starting point for all evaluations is the development of a contingency table. In Table 4-1, the preference of three BMPs is compared to resident type in a contingency table. In this case both categorical variables are nominal. In this example, 45 of the 102 residents that own the house they occupy used BMP_1. There were a total of 174 observations.

To test for independence, the sum of the squared differences between the expected (E_{ij}) and the observed (O_{ij}) count summed over all cells is computed as (Helsel and Hirsch, 1995)

$$\chi_{ct} = \sum_{i=1}^{m} \sum_{j=1}^{k} \frac{(O_{ij} - E_{ij})^2}{E_{ij}} \qquad (4\text{-}6)$$

where E_{ij} is equal to A_iC_j/N. χ_{ct} is compared to the $1\text{-}\alpha$ quantile of the χ^2 distribution with $(m\text{-}1)(k\text{-}1)$ degrees of freedom (see Table A3).

In the example presented in Table 4-1, the symbols listed in the parentheses correspond to the above equation. Note that *k* corresponds to the three types of BMPs and *m* corresponds to the three different types of residents. Table 4-2 shows computed values of E_{ij} and $(O_{ij}\text{-}E_{ij})^2/E_{ij})$ in parentheses for the example data. χ_{ct} is equal to 14.60. From Table A3, the 0.95 quantile of the χ^2 distribution with 4 degrees of freedom is 9.488. H_0 is rejected; the selection of BMP is not random among the different resident types. The largest values in the parentheses in Table 4-2 give an idea as to which combinations of resident type and BMP are noteworthy. In this example, it appears that BMP_2 is preferred to BMP_1 for those residents that rent the house they occupy.

Table 4-1. Contingency table of observed resident type and implemented BMP.

Resident Type	BMP_1	BMP_2	BMP_3	Row Total, A_i
Rent	10 (O_{11})	30 (O_{12})	17 (O_{13})	57 (A_1)
Own	45 (O_{21})	32 (O_{22})	25 (O_{23})	102 (A_2)
Seasonal	8 (O_{31})	3 (O_{32})	4 (O_{33})	15 (A_3)
Column Total, C_j	63 (C_1)	65 (C_2)	46 (C_3)	174 (N)

Key to Symbols:
O_{ij} = number of observations for the *i*th resident and *j*th BMP type
A_i = row total for the *i*th resident type (total number of observations for a given resident type)
C_j = column total for the *j*th BMP type (total number of observations for a given BMP type)
N = total number of observations

Table 4-2. Contingency table of expected resident type and implemented BMP. (Values in parentheses correspond to $(O_{ij}-E_{ij})^2/E_{ij}$.)

Resident Type	BMP_1	BMP_2	BMP_3	Row Total
Rent	20.64 (5.48)	21.29 (3.56)	15.07 (0.25)	57
Own	36.93 (1.76)	38.10 (0.98)	26.97 (0.14)	102
Seasonal	5.43 (1.22)	5.60 (1.21)	3.97 (0.00)	15
Column Total	63	65	46	174

Now consider that in addition to evaluating information regarding the resident and BMP type, we also recorded a value from 1 to 5 indicating how well the BMP was installed and maintained, with 5 indicating the best results. In this case, the BMP implementation rating is ordinal. Using the same notation as before, the average rank of observations in row x, R_x, is equal to (Helsel and Hirsch, 1995)

$$R_x = \sum_{i=1}^{x-1} A_i + (A_x + 1)/2 \qquad (4\text{-}7)$$

where A_i corresponds to the row total. The average rank of observations in column j, D_j, is equal to

$$D_j = \frac{\sum_{i=1}^{m} O_{ij} R_i}{C_j} \qquad (4\text{-}8)$$

where C_j corresponds to the column total. The Kruskal-Wallis test statistic is then computed as

$$K = (N - 1)\frac{\sum_{j=1}^{k} C_j D_j^2 - N\left[\frac{N+1}{N}\right]^2}{\sum_{i=1}^{m} A_i R_i^2 - N\left[\frac{N+1}{N}\right]^2} \qquad (4\text{-}9)$$

where K is compared to the χ^2 distribution with *k-1* degrees of freedom. This is the most general form of the Kruskal-Wallis test since it is a comparison of distribution shifts rather than shifts in the median (Helsel and Hirsch, 1995).

Table 4-3 is a continuation of the previous example indicating the BMP implementation rating for each BMP type. For example, 29 of the 70 observations that were given a rating of 4 are associated with BMP_2. The terms inside the parentheses of Table 4-3 correspond to the terms used in Equations 4-7 to 4-9. Note that k corresponds to the three types of BMPs and m corresponds to the five different levels of BMP implementation. Using Equation 4-9 for the data in Table 4-3, K is equal to 14.86. Comparing this value to 5.991 obtained from Table A3, there is a significant difference in the quality of implementation between the three BMPs.

The last type of categorical data evaluation considered in this chapter is when both variables are ordinal. The Kendall τ_b for tieddata can be used for this analysis. The statistic τ_b is calculated as (Helsel and Hirsch, 1995)

$$\tau_b = \frac{S}{\frac{1}{2}\sqrt{(N^2 - SS_a)(N^2 - SS_b)}} \qquad (4\text{-}10)$$

Table 4-3. Contingency table of implemented BMP and rating of installation and maintenance.

BMP Implementation Rating	BMP_1	BMP_2	BMP_3	Row Total, A_i
1	1 (O_{11})	2 (O_{12})	2 (O_{13})	5 (A_1)
2	7 (O_{21})	3 (O_{22})	5 (O_{23})	15 (A_2)
3	15 (O_{31})	16 (O_{32})	26 (O_{33})	57 (A_3)
4	32 (O_{41})	29 (O_{42})	9 (O_{43})	70 (A_4)
5	8 (O_{51})	15 (O_{52})	4 (O_{53})	27 (A_5)
Column Total, C_j	63 (C_1)	65 (C_2)	46 (C_3)	174 (N)

Key to Symbols:
O_{ij} = number of observations for the *i*th BMP implementation rating and *j*th BMP type
A_i = row total for the *i*th BMP implementation rating (total number of observations for a given BMP implementation rating)
C_j = column total for the *j*th BMP type (total number of observations for a given BMP type)
N = total number of observations

where S, SS_a, and SS_c are computed as

$$S = \sum_{all\,xy}\left[\sum_{i>x}\sum_{j>y} O_{xy}O_{ij} - \sum_{i<x}\sum_{j<}\right] \qquad (4\text{-}11)$$

$$SS_a = \sum_{i=1}^{m} A_i^2 \qquad (4\text{-}12)$$

$$SS_c = \sum_{j=1}^{k} C_j^2 \qquad (4\text{-}13)$$

To determine whether τ_b is significant, S is modified to a normal statistic using

$$Z_S = \begin{cases} \dfrac{S-1}{\sigma_S} & if\ S>0 \\ \dfrac{S-1}{\sigma_S} & if\ S<0 \end{cases} \qquad (4\text{-}14)$$

where

$$\sigma_s = \sqrt{\frac{N^3}{9}\left(1-\sum_{i=1}^{m} a_i^3\right)\left(1-\sum_{j=1}^{k}\right)} \qquad (4\text{-}15)$$

where Z_S is zero if S is zero. The values of a_i and c_i are computed as A_i/N and C_i/N, respectively.

Table 4-4 presents the BMP implementation ratings that were taken in three separate years. For example, 15 of the 57 observations that were given a rating of 3 are associated with Year 2. Using Equations 4-11 and 4-15, S and σ_s are equal to 2,509 and 679.75, respectively. Therefore, Z_s is equal to (2509-1)/679.75 or 3.69. Comparing this value to a value of 1.96 obtained from Table A1 ($\alpha/2$=0.025) indicates that BMP implementation is improving with time.

Table 4-4. Contingency table of implemented BMP and sample year.

BMP Implementation Rating	**Year 1**	**Year 2**	**Year 3**	**Row Total, A_i**	**a_i**
1	2 (O_{11})	1 (O_{12})	2 (O_{13})	5 (A_1)	0.029
2	5 (O_{21})	7 (O_{22})	3 (O_{23})	15 (A_2)	0.086
3	26 (O_{31})	15 (O_{32})	16 (O_{33})	57 (A_3)	0.328
4	9 (O_{41})	32 (O_{42})	29 (O_{43}).	70 (A_4)	0.402
5	4 (O_{51})	8 (O_{52})	15 (O_{53})	27 (A_5)	0.155
Column Total, C_j	46 (C_1)	63 (C_2)	65 (C_3)	174 (N)	
c_j	0.264	0.362	0.374		

Key to Symbols:
O_{ij} = number of observations for the *i*th BMP implementation rating and *j*th year
A_i = row total for the *i*th BMP implementation rating (total number of observations for a given BMP implementation rating)
C_j = column total for the *j*th BMP type (total number of observations for a given year)
N = total number of observations
a_i = A_i/N
c_j = C_j/N

CHAPTER 5. CONDUCTING THE EVALUATION

5.1 INTRODUCTION

This chapter addresses the process of determining whether urban MMs or BMPs are being implemented and whether they are being implemented according to approved standards or specifications. Guidance is provided on what should be measured to assess MM and BMP implementation, as well as methods for collecting the information, including physical site evaluations, mail- and/or telephone-based surveys, personal interviews, and aerial reconnaissance and photography. Designing survey instruments to avoid error and rating MM and BMP implementation are also discussed.

Evaluation methods are separated into two types: Expert evaluations and self-evaluations. Expert evaluations are those in which actual site investigations are conducted by trained personnel to gather information on MM or BMP implementation. Self-evaluations are those in which answers to a predesigned questionnaire or survey are provided by the person being surveyed, for example a local government representative or homeowner (see ***Example***). The answers provided are used as survey results. Self-evaluations might also include examination of materials related to a site, such as permit applications or inspection reports. Extreme caution should be exercised when using data from self-evaluations as the basis for assessing MM or BMP compliance since they are not typically reliable for this purpose (i.e., most people will not report failure or non-compliance). Each of these evaluation methods has advantages and disadvantages that should be considered prior to deciding which one to use or in what combination to use them. Aerial reconnaissance and photography can be used to support either evaluation method.

Self-evaluations are useful for collecting information on the level of awareness that residents, developers, or local government representatives of have of MMs or BMPs, dates of BMP implementation or inspection, soil conditions, which MMs or BMPs were implemented, and whether the assistance of a local or private BMP implementation professional was used. However, the type of or level of detail of information that can be obtained from self-evaluations might be inadequate to satisfy the objectives of a MM or BMP implementation survey. If this is the case, expert evaluations might be called for. Expert evaluations are necessary if the information on MM or BMP implementation that is required must be more detailed or more reliable than that that can be obtained with self-evaluations. Examples of information that would be obtained reliably only through an expert evaluation include an objective assessment of the adequacy of MM or BMP implementation, the degree to which site-specific factors (e.g., type of vegetative cover, soil type, or presence of a water body) influenced MM or BMP implementation, or the need for changes in standards and specifications for MM or BMP implementation. Sections 5.3 and 5.4 discuss expert evaluations and self-evaluations, respectively, in more detail.

A survey of lawn care practices in the Westmorland neighborhood of the city of Madison, Wisconsin was conducted by telephone interviews, after advance notice was sent to homeowners. The objectives of the survey were to:

- *Determine the number of people who fertilized their lawn either themselves or through a professional service.*
- *Identify the usage of fertilizer (i.e., when it was applied and the quantity applied).*
- *Determine the brands and types of fertilizers used.*
- *Identify the pattern of usage of separate weed killers and insecticides.*

The survey provided information on:

- *The percentage of homeowners that fertilized their lawns themselves.*
- *The demographic profile (e.g., sex, age, number of children) of the homeowners in the area that were most likely to use a professional service.*
- *The annual frequency of fertilizer applications.*
- *The type of equipment used for fertilizer applications.*
- *The percentage of homeowners who said they followed manufacturer recommendations for fertilizer applications.*
- *The percentage of homeowners who used fertilizer/insecticide combinations.*
- *The percentage of homeowners that used separate weed killers and insecticides.*
- *How the homeowners disposed of grass clippings.*

Example . . . *Survey of lawn care practices.* (Gene Kroupa & Associates, 1995)

Other important factors to consider when choosing variables include the time of year when the BMP compliance survey will be conducted and when BMPs were installed. Some urban BMPs, or aspects of their implementation that can be analyzed, vary with time of year, phase of construction, or length of time after having been installed. The temporary controls for erosion and sediment control, for instance, would not be inspected after construction is complete and a site has been stabilized. Variables that are appropriate to time-specific factors should be chosen. Concerning BMPs that have been in place for some time, the adequacy of implementation might be of less interest than the adequacy of the operation and maintenance of the BMP. For example, it might be of interest to inspect bridge runoff systems for proper cleaning and maintenance rather than to determine whether the number and spacing of runoff drains is sufficient for the particular bridge. If numerous BMPs are being analyzed during a single site visit, variables that relate to different aspects of BMP installation, operation, and maintenance might be chosen separately for each BMP to be inspected.

Aerial reconnaissance and photography is another means available for collecting information on urban or watershed practices, though many of the MMs and BMPs used in urban areas might be difficult if not impossible to identify on aerial photographs. Aerial reconnaissance and photography are discussed in detail in Section 5.5.

The general types of information obtainable with self-evaluations are listed in Table 5-1. Regardless of the approach(es) used, proper and thorough preparation for the evaluation is the key to success.

5.2 Choice of Variables

Once the objectives of a BMP implementation or compliance survey have been clearly defined, the most important factor in the assessment of MM or BMP implementation is the determination of which variable(s) to measure. A good variable provides a direct measure of how well a BMP was or is being implemented. Individual variables should provide measures of different factors related to BMP implementation. The best variables are those which are measures of the adequacy of MM or BMP implementation and are based on quantifiable expressions of conformance with state standards and specifications. As the variables that are used become less directly related to actual MM or BMP implementation, their accuracy as measures of BMP implementation decreases.

Examples of useful variables could include the change in the quantity of household hazardous waste collected or the percent of onsite disposal systems in a subwatershed that are operating properly, both of which would be expressed in terms of conformance with applicable state and/or local standards and specifications. Less useful variables measure factors that are related to MM and BMP implementation but do not necessarily provide an accurate measure of their implementation. Examples of these types of variables are the number of runoff conveyance structures constructed in a year and the number of onsite disposal systems approved for installation. Other poor variables would be the passage of legislation requiring MM or BMP application on construction sites, development of an public information program for lawn management, or the number of requests for information on household hazardous waste disposal. Although these variables relate to MM or BMP implementation, they provide no real information on whether MMs or BMPs are actually being implemented or whether they are being implemented properly.

Variables generally will not directly relate to MM implementation, as most urban MMs are combinations of several BMPs. Measures of MM implementation, therefore, usually will be based on separate assessments of two or more BMPs, and the implementation of each BMP will be based on a unique set of variables. Some examples of BMPs related to EPA's Site Development Management Measure, variables for assessing compliance with the BMPs, and related standards and specifications that might be required by local regulatory authorities are presented in Figure 5-1. Because developers and homeowners choose to implement or not implement MMs or BMPs based on site-specific conditions, it is also appropriate to apply varying weights to the variables chosen to assess MM and BMP implementation to correspond to site-specific conditions. For example, variables related to onsite disposal systems might be de-emphasized—and other, more applicable variables emphasized more—in areas where most homes are

Table 5-1. General types of information obtainable with self-evaluations and expert evaluations.

Information obtainable from Self-Evaluations

Background Information

- *Type of development installed (e.g., residential, commercial, industrial, recreational)*
- *Percent impervious area*
- *Inspection schedule*
- *Operation and maintenance practices*
- *Map*

Management Measures / Best Management Practices

- *Nonstructural practices*
- *BMPs installed*
- *Dates of MM / BMP installation*
- *Design specifications of BMPs*
- *Type of water body or area protected*
- *Previous management measures used*

ESC Plans (for construction)

- *Preparation of ESC plans*
- *Dates of plan preparation and revisions*
- *Date of initial plan implementation*
- *Total acreage under management*
- *Certification requirements*

Information that Requires Expert Evaluations

- *Design sufficiency*
- *Installation sufficiency*
- *Adequacy of operation / maintenance*
- *Confirmation of information from self-evaluations*

Site Development Management Measure

Plan, design, and develop sites to:

(1) Protect areas that provide important water quality benefits and/or are particularly susceptible to erosion and sediment loss;

(2) Limit increases of impervious areas, except where necessary;

(3) Limit land disturbance activities such as clearing and grading, and cut and fill to reduce erosion and sediment loss; and

(4) Limit disturbance of natural drainage features and vegetation.

Related BMPs, measurement variables, and standards and specifications:

Management Measure Practice	***Potential Measurement Variable***	***Example Related Standards and Specifications***
· Phasing and limiting areas of disturbance	*· Length of time disturbed area left without stabilization (temporary or permanent)*	*· Maximum time an area may be left unstabilized* *· Maximum area that may be disturbed at one time, depending on type of construction and project*
· Preserving natural drainage features and natural depressional storage areas	*· Degree to which postdevelopment landscape preserves predevelopment landscape features*	*· Site-specific requirements for preservation of natural drainage features, determined during the permitting process*
· Minimizing imperviousness	*· Percent impervious surface* *· Percent increase in impervious surface*	*· Maximum imperviousness, depending on type of development* *· Maximum percent increase in imperviousness, based on type of development*
· Minimum disturbance / minimum maintenance	*· Quantity of land altered by development from its predevelopment condition*	*· Guidelines for protection of natural vegetation and site characteristics, proposed for project during project development*

Figure 5-1. Potential variables and examples of implementation standards and specifications that might be useful for evaluating compliance with the New Development Management Measure.

connected to a sewer system. Similarly, on a construction site near a water body, variables related to sediment runoff and chemical deposition (pesticide use, fertilizer use) might be emphasized over other variables to arrive at a site-specific rating of the adequacy of MM or BMP implementation.

The purpose for which the information collected during a MM or BMP implementation survey will be used is another important consideration when selecting variables. An implementation survey can serve many purposes beyond the primary purpose of assessing MM and BMP implementation. For instance, variables might be selected to assess compliance with each category of BMP that is of interest and to assess overall compliance with BMP specification and standards. In addition, other variables might be selected to assess the effect that specific circumstances have on the ability or willingness of homeowners or developers to comply with BMP implementation standards or specifications. For example, the level of participation in a household hazardous waste collection program could be investigated with respect to variables for collection locations and hours of operation. The information obtained from evaluations using the latter type of variable could be useful for modifying MM or BMP implementation standards and specifications for application to particular types of developments or site conditions.

Table 5-2 provides examples of good and poor variables for the assessment of implementation of the urban MMs developed by EPA (USEPA, 1993a). The variables listed in the table are only examples, and local or regional conditions should ultimately dictate what variables should be used. The Center for Watershed Protection (CWP) published a report, *Environmental indicators to assess stormwater control programs and practices* (Clayton and Brown, 1996), that contains additional information on this subject. CWP also recommended that it might be necessary to evaluate BMP specifications to determine whether those for "older" structural BMPs are still appropriate for pollution prevention.

5.3 Expert Evaluations

5.3.1 Site Evaluations

Expert evaluations are the best way to collect reliable information on MM and BMP implementation. They involve a person or team of people visiting individual sites and speaking with homeowners and/or developers to obtain information on MM and BMP implementation (see Example). For many of the MMs, assessing and verifying compliance will require a site visit and evaluation. The following should be considered before expert evaluations are conducted:

- *Obtaining permission of the homeowner or developer.* Without proper authorization to visit a site from the homeowner or developer, the relationship between the regulated community and the local regulatory authority, and any future regulatory or compliance action, could be jeopardized.

- *The type(s) of expertise needed to assess proper implementation.* For some MMs, a team of trained personnel might be required to determine whether MMs have been implemented properly.

Table 5-2. Example variables for assessing management measure implementation.

Management Measure	Good Variable	Poor Variable	Appropriate Sampling Unit
URBAN RUNOFF			
New Development	· Number of county staff trained in ESC control. · Width of filter strips relative to area drained.	· Allocation of funding for development of education materials. · Scheduled frequency of runoff control maintenance.	· Subwatershed · Development site
Watershed Protection	· Percent of highly erodible soils left in an undeveloped state. · Percent natural drainage ways altered.	· Development of watershed analysis GIS system. · Assessed fines for violations of setback standards.	· Subwatershed
Site Development	· Ratio of area of land with structures to total disturbed land at a development site · Area of environmentally sensitive land to total area of same disturbed during construction	· Number of erosion and sediment control plans developed.	· Subwatershed
CONSTRUCTION ACTIVITIES			
Construction Site Erosion and Sediment Control (ESC)	· Distance runoff travels on disturbed soils before it is intercepted by a runoff control device (relative to slope and soil type). · Adequacy of ESC practices relative to soil type, slope, and precipitation.	· Number of ESC BMPs used at a construction site. · Number of ESC plans written.	· Development site
Construction Site Chemical Control	· Proper installation and use of designated area for chemical and petroleum product storage and handling. · Proper timing and application rate of nutrients at development site.	· Content and quality of spill prevention and control plan. · Number of approved nutrient management plans.	· Development site
EXISTING DEVELOPMENT			
Existing Development	· Proper operation and maintenance of surface water runoff management facilities. · Installation of appropriate BMPs in areas assigned priority as being in need of structural NPS controls.	· Development of a schedule for BMP implementation. · Setting priorities for structural improvements in development areas.	· Subwatershed

Table 5-2. (cont.)

Management Measure	Good Variable	Poor Variable	Appropriate Sampling Unit
ONSITE DISPOSAL SYSTEMS			
New Onsite Disposal Systems	· Proper siting and installation of new OSDS. · Density of development with OSDS in areas with nitrogen-limited waters	· Number of OSDS installed. · Reduction in garbage disposal sales.	· Subwatershed · City · Town
Operating Onsite Disposal Systems	· Increase in proper OSDS operation and maintenance 6 months after a public education campaign. · Average time between OSDS maintenance visits.	· Scheduled frequency of OSDS inspections. · Authorization of funding for public education campaign on OSDS.	· Subwatershed · City · Town
POLLUTION PREVENTION			
Pollution Prevention	· Increase in volume of household hazardous wastes collected. · Miles of roads adopted for citizen cleanup and volume of trash collected.	· Number of licenses issued to lawn care companies offering "chemical-free" lawn care. · Development of pollution prevention campaigns by nongovernmental organizations.	· City · Town
ROADS, HIGHWAYS, AND BRIDGES			
Planning, Siting, and Developing Roads and Highways	· Right-of-ways set aside for roads and highways based on projected future growth, and appropriateness of land set aside for such use.	· Miles of road constructed.	· Subwatershed
Bridges	· Total distance of bridges in environmentally sensitive areas.	· Number of bridges constructed.	· Subwatershed
Construction Projects	· Installation of ESC practices early in construction project. · ESC practices installed early in construction project.	· Number of ESC plans prepared and approved. · Number of ESC BMPs used during construction.	· Subwatershed

Table 5-2. (cont.)

Management Measure	Good Variable	Poor Variable	Appropriate Sampling Unit
ROADS, HIGHWAYS, AND BRIDGES (cont.)			
Construction Site Chemical Control	· Proper installation and use of designated area for chemical and petroleum product storage and handling. · Proper timing and application rate of nutrients at development site.	· Pounds of herbicide applied.	· Subwatershed
Operation and Maintenance	· Operating efficiency of NPS pollution control BMPs. · Ratio of exposed slopes and\or damaged vegetated areas to 100 m of roadway length. · Frequency of street sweeping.	· Purchase of salt application equipment.	· Subwatershed
Road, Highway, and Bridge Runoff Systems	· Adherence to schedule for implementation of runoff controls on roadways determined to need same. · Percent of roadway refurbishment projects that include runoff control improvements on roads needing same.	· Purchase of land for location of treatment facilities.	· Subwatershed

In Delaware, private construction site inspectors make at least weekly site visits to large or significant construction sites. The private inspectors are trained by the state and report violations of ESC regulations and inadequacies in ESC plans or BMP implementation to the state or local ESC agency, the developer, and the contractor. They also offer timely on-site technical assistance. While not a comprehensive ESC BMP implementation inventory program, it can be used as a model for the development of such a program.

***Example . . .** Delaware construction site reviews.*

- *The activities that should occur during an expert evaluation.* This information is necessary for proper and complete preparation for the site visit, so that it can be completed in a single visit and at the proper time.

- *Inspection reports or certifications (developed during construction or as the result of other studies) might exist for some BMPs.* The team of trained personnel should consider whether the BMP was built to standards that a "new" BMP would be built to meet the MMs. (This might require reviewing the engineering design and specifications.) If the standards are comparable, then a previous inspection report or certification might be acceptable in lieu of a detailed site visit and evaluation.

- *The method of rating the MMs and BMPs.* MM and BMP rating systems are discussed below.

- *Consistency among evaluation teams and between site evaluations.* Proper training and preparation of expert evaluation team members are crucial to ensure accuracy and consistency.

- *The collection of information while at a site.* Information collection should be facilitated with preparation of data collection forms that include any necessary MM and BMP rating information needed by the evaluation team members.

- *The content and format of post-evaluation discussions.* Site evaluation team members should bear in mind the value of postevaluation discussion among team members. Notes can be taken during the evaluation concerning any items that would benefit from group discussion.

Evaluators might consist of a single person suitably trained in urban expert evaluation to a group of professionals with varied expertise. The composition of evaluation teams will depend on the types of MMs or BMPs being evaluated. Potential team members could include:

- Civil engineer
- Land use planner
- Hydrologist
- Soil scientist
- Water quality expert

The composition of evaluation teams can vary depending on the purpose of the evaluation, available staff and other resources, and the geographic area being covered. All team members should be familiar with the required MMs and BMPs, and each team should have a member who has previously participated in an expert evaluation. This will ensure familiarity with the technical aspects of the MMs and BMPs that will be rated during the evaluation and the expert evaluation process.

Training might be necessary to bring all team members to the level of proficiency needed to conduct the expert evaluations. State or local regulatory personnel should be familiar with urban conditions, state BMP standards and specifications, and proper BMP implementation, and therefore are generally well qualified to teach these topics to evaluation team members who are less familiar with them. Local regulatory agency representatives or other specialists who have participated in BMP implementation surveys might be enlisted to train evaluation team members about the actual conduct of expert evaluations. This training should include identification of BMPs particularly critical to water quality protection, analysis of erosion potential, and other aspects of BMP implementation that require professional judgement, as well as any standard methods for measurements to judge BMP implementation against state standards and specifications.

Alternatively, if only one or two individuals will be conducting expert evaluations, their training in the various specialties, such as those listed above, necessary to evaluate the quality of MM and BMP implementation could be provided by a team of specialists who are familiar with urban BMPs and nonpoint source pollution.

In the interest of consistency among the evaluations and among team members, it is advisable that one or more mock evaluations take place prior to visiting selected sample sites. These "practice sessions" provide team members with an opportunity to become familiar with MMs and BMPs as they should be implemented under different site conditions, gain familiarity with the evaluation forms and meanings of the terms and questions on them, and learn from other team members with different expertise. Mock evaluations are valuable for ensuring that all evaluators have a similar understanding of the intent of the questions, especially for questions whose responses involve a degree of subjectivity on the part of the evaluator.

Where expert evaluation teams are composed of more than two or three people, it might be helpful to divide up the various responsibilities for conducting the expert evaluations among team members ahead of time to avoid confusion at the site and to be certain that all tasks are completed but not duplicated. Having a spokesperson for the group who will be responsible for communicating with the homeowner or developer—prior to the expert evaluation, at the expert evaluation if they are present, and afterward—might also be helpful. A local regulatory agency representative is generally a good choice as spokesperson because he/she can represent the county or municipal authorities. Newly-formed evaluation teams might benefit most from a division of labor and selection of a team leader or team coordinator with experience in expert evaluations who will be responsible for the quality of the expert evaluations. Smaller teams and larger teams that have experience working together might find that a division of responsibilities is not necessary. If responsibilities are to be assigned, mock evaluations can be a good time to work out these details.

5.3.2 Rating Implementation of Management Measures and Best Management Practices

Many factors influence the implementation of MMs and BMPs, so it is sometimes necessary to use best professional judgment (BPJ) to rate their implementation and BPJ will almost always be necessary when rating overall BMP compliance at a site. Site-specific factors such as soil type, amount of area exposed, and topography affect the implementation of erosion and sediment control BMPs, for instance, and must be taken into account by evaluators when rating MM or BMP implementation. Implementation of MMs will often be based on implementation of more than one BMP, and this makes rating MM implementation similar to rating overall BMP implementation at a site. Determining an overall rating involves grouping the ratings of implementation of individual BMPs into a single rating, which introduces more subjectivity than rating the implementation of individual BMPs based on standards and specifications. Choice of a rating system and rating terms, which are aspects of proper evaluation design, is therefore important in minimizing the level of subjectivity associated with overall BMP compliance and MM implementation ratings. When creating overall ratings, it is still important to record the detailed ratings of individual BMPs as supporting information.

Individual BMPs, overall BMP compliance, and MMs can be rated using a binary approach (e.g., pass/fail, compliant/ noncompliant, or yes/no) or on a scale with more than two choices, such as 1 to 5 or 1 to 10 (where 1 is the worst—see ***Example***). The simplest method of rating MM and BMP implementation is the use of a binary approach. Using a binary approach, either an entire site or individual MMs or BMPs are rated as being in compliance or not in compliance with respect to specified criteria. Scale systems can take the form of ratings from poor to excellent, inadequate to adequate, low to high, 1 to 3, 1 to 5, and so forth.

A possible rating scale from 1 to 5 might be:

- *5 = Implementation exceeds requirements*
- *4 = Implementation meets requirements*
- *3 = Implementation has a minor departure from requirements*
- *2 = Implementation has a major departure from requirements*
- *1 = Implementation is in gross neglect of requirements*

where:

Minor departure is defined as "small in magnitude or localized," major departure is defined as "significant magnitude or where the BMPs are consistently neglected" and gross neglect is defined as "potential risk to water resources is significant and there is no evidence that any attempt is made to implement the BMP."

Example... of a rating scale (adapted from Rossman and Phillips, 1992).

Whatever form of scale is used, the factors that would individually or collectively qualify a site, MM, or BMP for one of the rankings should be clearly stated. The more choices that are added to the scale, the smaller and smaller the difference between them becomes and each must therefore be defined more specifically and accurately. This is especially important if different teams or individuals rate sites separately. Consistency among the ratings then depends on each team or individual evaluator knowing precisely what the criteria for each rating option mean. Clear and precise explanations of the rating scale can also help avoid or reduce disagreements among team members. This applies equally to a binary approach. The factors, individually or collectively, that would cause a site, MM, or BMP to be rated as not being in compliance with design specifications should be clearly stated on the evaluation form or in support documentation.

Rating sites or MMs and BMPs on a scale requires a greater degree of analysis by the evaluation team than does using a binary approach. Each higher number represents a better level of MM or BMP implementation. In effect, a binary rating approach is a scale with two choices; a scale of low, medium, and high (compliance) is a scale with three choices. Use of a scale system with more than two rating choices can provide more information to program managers than a binary rating approach, and this factor must be weighted against the greater complexity involved in using one. For instance, a survey that uses a scale of 1 to 5 might result in one MM with a ranking of 1, five with a ranking of 2, six with a ranking of 3, eight with a ranking of 4, and

Case Study: Field Test of a Randomized BMP Sampling Design

Maryland Environmental Service (MES), an organization under contract to the St. Mary's County, Maryland, Department of Public Works to inventory and inspect all storm water BMPs in the county, conducted a field test of the sampling techniques in this guidance to determine how well an inspection of a sample of storm water BMPs could predict the condition of all storm water BMPs in the county (MES, 2000). To do this, MES first conducted its annual inventory and inspection of 30 storm water BMPs, or approximately one-third of the county's total. The BMPs, which included a variety of types of structural storm water control facilities, were inspected for deficiencies in 38 areas in accordance with Maryland's inspection requirements. The inspectors noted any maintenance requirements for the BMPs as well as their condition and function.

From the results, the total number of areas found to have deficiencies from all 30 BMPs inspected, as well as the percentage of the BMPs inspected that were deficient in each of the 38 areas was tallied. Then, MES randomly selected 10 of the 30 storm water BMPs that they had inspected to determine how well the same analysis performed on these 10 facilities would compare to the analysis of the 30 facilities. The results are shown in the table below.

	BMPs had deficiencies in:	BMPs with individual deficiencies ranged from:
Analysis of all 30 BMPs	20 of 38 areas	3 to 30 percent
Analysis of a sample of 10 BMPs	7 of 10 areas	10 to 40 percent

Since the 30 BMPs inspected included a variety of types of BMPs, MES also analyzed a subsample of the BMPs inspected by including only BMPs with detention weirs in another analysis. Of the 30 BMPs inspected, 24 had detention weirs and were included in the second analysis. Ten of these were selected for the random sample within the group of 24. MES compared overall inspector ratings for the two groups. The results are shown in the table below. As can be seen the average inspector rating for all 24 BMPs with detention weirs easily falls within the 90 percent confidence interval associated with the average inspector rating associated with the sample of 10 BMPs.

	Average inspector rating was:
Analysis of all 24 BMPs with Detention Weirs	7.92
Analysis of a sample of 10 BMPs from the 24 (± 90% confidence interval)	7.6±0.9

Lesson: A sample of BMPs can be used to predict the condition of a larger group of BMPs, but results are more reliable if the random sample design can be used to eliminate sources of error that could result in erroneous interpretation of the results.

five with a ranking of 5. Precise criteria would have to be developed to be able to ensure consistency within and between survey teams in rating the MMs, but the information that only one MM was implemented poorly, 11 were implemented below standards, 13 met or were above standards, and 5 were implemented very well might be more valuable than the information that 18 MMs were found to be in compliance with design specifications, which is the only information that would be obtained with a binary rating approach.

If a rating system with more than two ratings is used to collect data, the data can be analyzed either by using the original rating data or by first transforming the data into a binomial (i.e., two-choice rating) system. For instance, ratings of 1 through 5 could be reduced to two ratings by grouping the 1s, 2s, and 3s together into one group (e.g., inadequate) and the 4s and 5s into a separate group (e.g., adequate). If this approach is used, it is best to retain the original rating data for the detailed information they contain and to reduce the data to a binomial system only for the purpose of statistical analysis. Chapter 4, Section 4.5, contains information on the analysis of categorical data.

5.3.3 Rating Terms

The choice of rating terms used on the evaluation forms is an important factor in ensuring consistency and reducing bias, and the terms used to describe and define the rating options should be as objective as possible. For a rating system with a large number of options, the meanings of each option should be clearly defined. It is suggested to avoid using terms such as "major" and "minor" when describing erosion or pollution effects or deviations from prescribed MM or BMP implementation criteria, or to provide clear definitions for them in the context of the evaluation, because they might have different connotations for different evaluation team members. It is easier for an evaluation team to agree upon meaning if options are described in terms of measurable criteria and examples are provided to clarify the intended meaning. It is also suggested not to use terms that carry negative connotations. Evaluators might be disinclined to rate a MM or BMP as having a "major deviation" from an implementation criterion, even if justified, because of the negative connotation carried by the term. Rather than using such a term, observable conditions or effects of the quality of implementation can be listed and specific ratings (e.g., 1-5 or compliant/noncompliant for the criterion) can be associated with the conditions or effects. For example, instead of rating a stormwater management pond as having a "major deficiency," a specific deficiency could be described and ascribed an associated rating (e.g., "Structure is designed for no more than 5-hour attenuation of urban runoff = noncompliant").

Evaluation team members will often have to take specific notes on sites, MMs, or BMPs during the evaluation, either to justify the ratings they have ascribed to variables or for discussion with other team members after the survey. When recording notes about the sites, MMs, or BMPs, evaluation team members should be as specific as the criteria for the ratings. A rating recorded as "MM deviates highly from implementation criteria" is highly subjective and loses specific meaning when read by anyone other than the person who wrote the note. Notes should therefore be as objective and specific as possible.

An overall site rating is useful for summarizing information in reports, identifying the level of implementation of MMs and BMPs, indicating the likelihood that environmental protection is being achieved, identifying additional training or education needs; and conveying information to program managers, who are often not familiar with MMs or BMPs. For the purposes of preserving the valuable information contained in the original ratings of sites, MMs, or BMPs, however, overall ratings should summarize, not replace, the original data. Analysis of year-to-year variations in MM or BMP implementation, the factors involved in MM or BMP program implementation, and factors that could improve MM or BMP implementation and MM or BMP program success are only possible if the original, detailed site, MM, or BMP data are used.

Approaches commonly used for determining final BMP implementation ratings include calculating a percentage based on individual BMP ratings, consensus, compilation of aggregate scores by an objective party, voting, and voting only where consensus on a site or MM or BMP rating cannot be reached. Not all systems for arriving at final ratings are applicable to all circumstances.

5.3.4 Consistency Issues

Consistency among evaluators and between evaluations is important, and because of the potential for subjectivity to play a role in expert evaluations, consistency should be thoroughly addressed in the quality assurance and quality control (QA/QC) aspects of planning and conducting an implementation survey. Consistency arises as a QA/QC concern in the planning phase of an implementation survey in the choice of evaluators, the selection of the size of evaluation teams, and in evaluator training. It arises as a QA/QC concern while conducting an implementation survey in whether evaluations are conducted by individuals or teams, how MM and BMP implementation on individual sites is documented, how evaluation team discussions of issues are conducted, how problems are resolved, and how individual MMs and BMPs or whole sites are rated.

Consistency is likely to be best if only one to two evaluators conduct the expert evaluations and the same individuals conduct all of the evaluations. If, for statistical purposes, many sites (e.g., 100 or more) need to be evaluated, use of only one to two evaluators might also be the most efficient approach. In this case, having a team of evaluators revisit a subsample of the sites that were originally evaluated by one to two individuals might be useful for quality control purposes.

If teams of evaluators conduct the evaluations, consistency can be achieved by keeping the membership of the teams constant. Differences of opinion, which are likely to arise among team members, can be settled through discussions held during evaluations, and the experience of team members who have done past evaluations can help guide decisions. Preevaluation training sessions, such as the mock evaluations discussed above, will help ensure that the first few expert evaluations are not "learning" experiences to such an extent that those sites must be revisited to ensure that they receive the same level of scrutiny as sites evaluated later.

If different sites are visited by different teams of evaluators or if individual evaluators are assigned to different sites, it is especially important that consistency be established before the evaluations are conducted. For best results, discussions among evaluators should be held periodically during the evaluations to discuss any potential problems. For instance, evaluators could visit some sites together at the beginning of the evaluations to promote consistency in ratings, followed by expert evaluations conducted by individual evaluators. Then, after a few site or MM evaluations, evaluators could gather again to discuss results and to share any knowledge gained to ensure continued consistency.

As mentioned above, consistency can be established during mock evaluations held before the actual evaluations begin. These mock evaluations are excellent opportunities for evaluators to discuss the meaning of terms on rating forms, differences between rating criteria, and differences of opinion about proper MM or BMP implementation. A member of the evaluation team should be able to represent the state's position on the definition of terms and clarify areas of confusion.

Descriptions of MMs and BMPs should be detailed enough to support any ratings given to individual features and to the MM or BMP overall. Sketching a diagram of the MM or BMP helps identify design problems, promotes careful evaluation of all features, and provides a record of the MM or BMP for future reference. A diagram is also valuable when discussing the MM or BMP with the homeowner or developer or identifying features in need of improvement or alteration. Homeowners or developers can also use a copy of the diagram and evaluation when discussing their operations with local or state regulatory personnel. Photographs of MM or BMP features are a valuable reference material and should be used whenever an evaluator feels that a written description or a diagram could be inadequate. Photographs of what constitutes both good and poor MM or BMP implementation are valuable for explanatory and educational purposes; for example, for presentations to managers and the public.

5.3.5 Postevaluation Onsite Activities

It is important to complete all pertinent tasks as soon as possible after the completion of an expert evaluation to avoid extra work later and to reduce the chances of introducing error attributable to inaccurate or incomplete memory or confusion. All evaluation forms for each site should be filled out completely before leaving the site. Information not filled in at the beginning of the evaluation can be obtained from the site owner or developer if necessary. Any questions that evaluators had about the MMs and BMPs during the evaluation can be discussed, notes written during the evaluation can be shared and used to help clarify details of the evaluation process and ratings. The opportunity to revisit the site will still exist if there are points that cannot be agreed upon among evaluation team members.

Also, while the evaluation team is still on the site, the site owner or developer should be informed about what will follow; for instance, whether he/she will receive a copy of the report, when to expect it, what the results mean, and his/her responsibility in light of the evaluation, if any. Immediately following the evaluation is also an excellent time to discuss the findings with the site owner or developer if he/she was not present during the evaluation.

5.4 Self-Evaluations

5.4.1 Methods

Self-evaluations, while often not a reliable source of MM or BMP implementation data, can be used to augment data collected through expert evaluations or in place of expert evaluations where the latter cannot be conducted. In some cases, local or state regulatory personnel might have been involved directly with BMP selection and implementation and will be a source of useful information even if an expert evaluation is not conducted. Self-evaluations are an appropriate survey method for obtaining background information from homeowners or developers or other persons associated with BMP installation, such as contractors.

Mail, telephone, and mail with telephone follow-up are common self-evaluation methods (see ***Example***). Mail and telephone surveys are useful for collecting general information, such as the management measures that should be implemented on specific urban land types. Local regulatory agency personnel, county or municipal planning staff, or other state or local BMP implementation experts can be interviewed or sent a questionnaire that

The Center for Watershed Protection in Silver Spring, Maryland conducted a mail survey of erosion and sediment control (ESC) programs for small (< 5 acres) construction sites. The survey was sent to 219 jurisdictions located in all EPA regions and CWP received a 52% (113 surveys) response rate from the survey. The main objective of the survey was to identify innovative and effective ESC programs.

Through the mail survey, information was collected on the following:

- *The age of each program.*
- *Each program's requirements for permits (i.e., whether a separate process or part of the site development process).*
- *The applicability of permit requirements (i.e., whether applicability was based on site size or other criteria).*
- *The necessary conditions under which permit wavers would be issued.*
- *Whether the requirement for a permit was determined on a case-by-case basis, or whether certain aspects of the development (e.g., proximity to sensitive areas) would make obtaining a permit necessary.*
- *The size of populations in jurisdictions with ESC programs.*
- *Whether the ESC programs were mandated or voluntary.*
- *The level of detail required in ESC plans.*
- *Which ESC practices were used commonly.*
- *Who the enforcement agency was.*
- *What penalties could be imposed for non-compliance.*
- *A list of construction-related water quality problems common at small sites.*

Example . . . *Mail survey of ESC programs.* (Ohrel, 1996) requests very specific information. Recent advances in and increasing access to electronic means of communication (i.e., e-mail and the Internet) might make these viable survey instruments in the future.

To ensure comparability of results, information that is collected as part of a self-evaluation—whether collected through the mail, over the phone, or during site visits—should be collected in a manner that does not favor one method over the others. Ideally, telephone follow-up and on-site interviews should consist of no more than reading the questions on the questionnaire, without providing any additional explanation or information that would not have been available to those who responded through the mail. This approach eliminates as much as possible any bias associated with the different means of collecting the information. Figure 5-2 presents questions from a residential questionnaire developed for Prince George's County, Maryland, to determine residential "good housekeeping" practices. Questionnaire design is discussed in Section 5.4.3. It is important that the accuracy of information received through mail and phone surveys be checked. Inaccurate or incomplete responses to questions on mail and/or telephone surveys commonly result from survey respondents misinterpreting questions and thus providing misleading information, not including all relevant information in their responses, not wanting to provide some types of information, or deliberately providing some inaccurate responses. Therefore, the accuracy of information received through mail and phone surveys should be checked by selecting a subsample of the homeowners or other persons surveyed and conducting follow-up site visits.

5.4.2 Cost

Cost can be an important consideration when selecting an evaluation method. Site visits can cost several hundred dollars per site visited, depending on the type of inspection involved, the information to be collected, and the number of evaluators involved. Mail and/or telephone surveys can be an inexpensive means of collecting information, but their cost must be balanced with the type and accuracy of information that can be collected through them. Other costs also need to be figured into the overall cost of mail and/or telephone surveys, including follow-up phone calls and site visits to make up for a poor response to mailings and for accuracy checks. Additionally, the cost of questionnaire design must be considered, as a well-designed questionnaire is extremely important to the success of self-evaluations. Questionnaire design is discussed in the next section.

The number of evaluators used for site visits has an obvious impact on the cost of a MM or BMP implementation survey. Survey costs can be minimized by having one or two evaluators visit sites instead of having multiple-person teams visit each site. If the expertise of many specialists is desired, it might be cost-effective to have multiple-person teams check the quality of evaluations conducted by one or two evaluators. This can usually be done at a subsample of sites after they have been surveyed.

An important factor to consider when determining the number of evaluators to include on site visitation teams, and how to balance the use of one to two evaluators versus multiple-person teams, is the objectives of the

Questions about water quality and community activity factors:

1. Do you believe that rainwater runoff from streets, driveways, and parking lots causes water pollution in nearby streams?
Y ____ N ____ Don't know ____

2. Do you know where to report water pollution problems? Y ____ N ____

3. Do you know whether water from your lawn goes into Chesapeake Bay?
Y ____ N ____ Don't know ____

4. Do the storm drains in your neighborhood have "Drains into the Chesapeake Bay" stenciled on them? Y ____ N ____ Don't know ____

5. Please rank the following community issues according to their level of importance (1 = very important, 10 = not important)
____ Keeping trash and litter from accumulating in neighborhoods, parking lots, on main streets, and in commercial areas
____ Appearance and good maintenance of residential neighborhoods and commercial facilities
____ Organized youth programs
____ Protecting the environment (clean air and drinking water)
____ Stable property values
____ Crime
____ Having clean parks and recreational facilities
____ Traffic congestion on main roadways
____ Water pollution (polluted streams and waterways)
____ Other, please specify ____________________________________

Questions about lawn and garden maintenance activities:

1. Does a professional lawn care company fertilize your lawn? Y ____ N ____
(If yes, please proceed to question #5)

2. Please identify when and how often you fertilize your lawn:

Times per season	***Spring***	***Summer***	***Autumn***	***Winter***
Once	_____	_____	_____	_____
Twice	_____	_____	_____	_____
Three	_____	_____	_____	_____
Other	_____	_____	_____	_____

3. Indicate, to the best of your knowledge, how much fertilizer you use.
____ According to the instructions on the bag
____ pounds of nitrogen per 1,000 square feet
____ pounds of fertilizer per application
____ Don't know

Figure 5-2. Sample draft survey for residential "good housekeeping" practice implementation.

4. *Circle the pesticide/insecticide treatments that you perform on your lawn and/or garden.*

Insects:	*Spring*	*Summer*	*Autumn*	*Winter*	*Never*
Weeds:	*Spring*	*Summer*	*Autumn*	*Winter*	*Never*
Fungi:	*Spring*	*Summer*	*Autumn*	*Winter*	*Never*

5. *Please indicate how you dispose of your yard debris. (Mark all that apply)*
   ____ *Compost in backyard* ____ *Curbside trash pick-up*
   ____ *County/town composting program* ____ *Take off property to vacant lot or open space*
   ____ *Other: please specify* ________________________________

6. *Please use the space below to provide comments that you may have regarding yard maintenance and practices that may affect water quality.*

Questions about personal vehicle maintenance:

1. *Do you know how to report abandoned vehicles? Y ____ N ____*

2. *Please identify the number and age of vehicles that you currently own or lease:*

Number of vehicles	***Year***
________	*Pre-1980*
________	*1980 - 1990*
________	*1990 - present*

3. *Do you perform minor repairs or maintenance on your vehicle(s) at home?*
   Y ____ N ____

4. *How often do you change the oil in your vehicles at home?*
   ___ *monthly* ___ *quarterly* ___ *never*
   ___ *twice a year* ___ *once a year* ___ *don't know*

5. *How often do you change the antifreeze in your vehicle(s) at home?*
   ___ *twice a year* ___ *once a year*
   ___ *never* ___ *don't know*

6. *If you perform minor repairs or maintenance on your vehicle(s), please indicate where you dispose of the items listed below:*

	on ground	***in storm drain***	***gas station***	***home trash***	***other***
Engine oil	___	___	___	___	___
Antifreeze	___	___	___	___	___
Oil filters	___	___	___	___	___
Car batteries	___	___	___	___	___
Tires	___	___	___	___	___

Figure 5-2. (cont.)

survey. Cost notwithstanding, the teams conducting the expert evaluations must be sufficient to meet the objectives of the survey, and if the required teams would be too costly, then the objectives of the survey might need to be modified.

Another factor that contributes to the cost of a MM or BMP implementation survey is the number of sites to be surveyed. Once again, a balance must be reached between cost, the objectives of the survey, and the number of sites to be evaluated. Generally, once the objectives of the study have been specified, the number of sites to be evaluated can be determined statistically to meet required data quality objectives. If the number of sites that is determined in this way would be too costly, then it would be necessary to modify the study objectives or the data quality objectives. Statistical determination of the number of sites to evaluate is discussed in Section 3.3.

5.4.3 Questionnaire Design

Many books have been written on the design of data collection forms and questionnaires (e.g., Churchill, 1983; Ferber et al., 1964; Tull and Hawkins, 1990), and these can provide good advice for the creation of simple questionnaires that will be used for a single survey. However, for complex questionnaires or ones that will be used for initial surveys as part of a series of surveys (i.e., trend analysis), it is strongly advised that a professional in questionnaire design be consulted. This is because while it might seem that designing a questionnaire is a simple task, small details such as the order of questions, the selection of one word or phrase over a similar one, and the tone of the questions can significantly affect survey results. A professionally-designed questionnaire can yield information beyond that contained in the responses to the questions themselves, while a poorly-designed questionnaire can invalidate the results.

The objective of a questionnaire, which should be closely related to the objectives of the survey, should be extremely well thought out prior to its being designed. Questionnaires should also be designed at the same time as the information to be collected is selected to ensure that the questions address the objectives as precisely as possible. Conducting these activities simultaneously also provides immediate feedback on the attainability of the objectives and the detail of information that can be collected. For example, an investigator might want information on the extent of proper operation and maintenance of BMPs but might discover while designing the questionnaire that the desired information could not be obtained through the use of a questionnaire, or that the information that could be collected would be insufficient to fully address the chosen objectives. In such a situation the investigator could revise the objectives and questions before going further with questionnaire design.

Tull and Hawkins (1990) identified seven major elements of questionnaire construction:

1. Preliminary decisions
2. Question content
3. Question wording
4. Response format
5. Question sequence
6. Physical characteristics of the questionnaire
7. Pretest and revision.

Preliminary decisions include determining exactly what type of information is required, determining the target audience, and selecting the method of communication (e.g, mail, telephone, site visit). These subjects are addressed in other sections of this guidance.

The second step is to determine the content of the questions. Each question should generate one or more of the information requirements identified in the preliminary decisions. The ability of the question to elicit the necessary data needs to be assessed. "Double-barreled" questions, in which two or more questions are asked as one, should be avoided. Questions that require the respondent to aggregate several sources of information should be subdivided into several specific questions or parts. The ability of the respondent to answer accurately should also be considered when preparing questions. Some respondents might be unfamiliar with the type of information requested or the terminology used. Or a respondent might have forgotten some of the information of interest, or might be unable to verbalize an answer. Consideration should be given to the willingness of respondents to answer the questions accurately. If a respondent feels that a particular answer might be embarrassing or personally harmful, (e.g., might lead to fines or increased regulation), he or she might refuse to answer the question or might deliberately provide inaccurate information. For this reason, answers to questions that might lead to such responses should be checked for accuracy whenever possible.

The next step is the specific phrasing of the questions. Simple, easily understood language is preferred. The wording should not bias the answer or be too subjective. For instance, a question should not ask whether the local government adequately maintains structural BMPs (the likelihood of getting a negative response is low). Instead, a series of questions could ask whether the local government is responsible for operation and maintenance (O&M) of structural BMPs, how much staff and financial resources are dedicated to O&M, the frequency of inspection and maintenance, and the procedure for repair, if repair is necessary. These questions all request factual information of which the appropriate local government representative should be knowledgeable and they progress from simple to more complex. All alternatives and assumptions should be clearly stated on the questionnaire, and the respondent's frame of reference should be considered.

Fourth, the type of response format should be selected. Various types of information can best be obtained using open-ended, multiple-choice, or dichotomous questions. An open-ended question allows respondents to answer in any way they feel is appropriate. Multiple-choice questions tend to reduce some types of bias and are easier to tabulate and analyze; however, good multiple-choice questions can be more difficult to formulate. Dichotomous questions allow only two responses, such as "yes-no" or "agree-disagree." Dichotomous questions are suitable for determining points of fact, but must be very precisely stated and unequivocally solicit only a single piece of information.

The fifth step in questionnaire design is the ordering of the questions. The first questions should be simple to answer, objective, and interesting in order to relax the respondent. The questionnaire should move from topic to topic in a logical manner without confusing the respondent. Early questions that could bias the respondent should be avoided. There is evidence that response quality declines near the end of a long questionnaire (Tull and Hawkins, 1990). Therefore, more important information should be solicited early. Before presenting the questions, the questionnaire should explain how long (on average) it will take to complete and the types of information that will be solicited. The questionnaire should not present the respondent with any surprises.

The layout of the questionnaire should make it easy to use and should minimize recording mistakes. The layout should clearly show the respondent all possible answers. For mail surveys, a pleasant appearance is important for securing cooperation.

The final step in the design of a questionnaire is the pretest and possible revision. A questionnaire should always be pretested with members of the target audience. This will preclude expending large amounts of effort and then discovering that the questionnaire produces biased or incomplete information.

5.5 Aerial Reconnaissance and Photography

Aerial reconnaissance and photography can be useful tools for gathering physical site information quickly and comparatively inexpensively, and they are used in conservation for a variety of purposes. Aerial photography has been proven to be helpful for agricultural conservation practice identification (Pelletier and Griffin, 1988); rangeland monitoring (BLM, 1991); terrain stratification, inventory site identification, planning, and monitoring in mountainous regions (Hetzel, 1988; Born and Van Hooser, 1988); as well as for forest regeneration assessment (Hall and Alred, 1992) and forest inventory and analysis (Hackett, 1988). Factors such as the characteristics of what is being monitored, scale, and camera format determine how useful aerial photography can be for a particular purpose. For the purposes of urban area BMP implementation tracking, aerial photography could be a valuable tool for collecting information on a watershed, subwatershed, or smaller scale. For example, it could be useful to assess the condition of riparian vegetation, level of imperviousness in a subwatershed, or quantity and location of active construction sites in a specific area.

Pelletier and Griffin (1988) investigated the use of aerial photography for the identification of agriculture conservation practices. They found that practices that occupy a large area and have an identifiable pattern, such as contour cropping, strip cropping, terraces, and windbreaks, were readily identified even at a small scale (1:80,000) but that smaller, single-unit practices, such as sediment basins and sediment diversions, were difficult to identify at a small scale. They estimated that 29 percent of practices could be identified at a scale of 1:80,000, 45 percent could be identified at 1:30,000, 70 percent could be identified at 1:15,000, and over 90 percent could be identified at a scale of 1:10,000.

Camera format is a factor that also must be considered. Large-format cameras are generally preferred over small-format cameras (e.g., 35 mm), but are more costly to purchase and operate. The large negative size (9 cm x 9 cm) produced using a large-format camera provides the resolution and detail necessary for accurate photo interpretation. Large-format cameras can be used from higher altitudes than small-format cameras, and the image area covered by a large-format image at a given scale (e.g., 1:1,500) is much larger than the image area captured by a small-format camera at the same scale. Small-scale cameras (i.e., 35 mm) can be used for identifications that involve large-scale features, such as riparian areas and the extent of cleared land, and they are less costly to purchase and use than large-format cameras, but they are limited in the altitude that the photographs can be taken from and the resolution that they provide when enlarged (Owens, 1988).

BLM recommends the use of a large-format camera because the images provide the photo interpreter with more geographical reference points, it provides flexibility to increase sample plot size, and it permits modest navigational errors during overflight (BLM, 1991). Also, if hiring someone to take the photographs, most photo contractors will have large-format equipment for the purpose.

A drawback to the use of aerial photography is that urban BMPs that do not meet implementation or operational standards but that are similar to BMPs that do are indistinguishable from the latter in an aerial photograph (Pelletier and Griffin, 1988). Also, practices that are defined by managerial concepts rather than physical criteria, such as construction site chemical control or nutrient application rate, cannot be detected with aerial photographs.

Regardless of scale, format, or item being monitored, it is useful for photo interpreters to receive 2-3 days of training on the basic fundamentals of photo interpretation and that they be thoroughly familiar with the areas where the photographs that they will be interpreting were taken (BLM, 1991). A visit to the areas in photograph is recommended to improve correlation between the interpretation and actual site characteristics. Generally, after a few visits and interpretations of photographs of those areas, photo interpreters will be familiar with the photographic characteristics of the areas and the site visits can be reserved for verification of items in doubt.

Information on obtaining aerial photographs is available from the Natural Resources Conservation Service. Contact the Natural Resources Conservation Service at: NRCS National Cartography and Geospatial Center, Fort Worth Federal Center, Bldg 23, Room 60, P.O. Box 6567, Fort Worth, TX 76115-0567; 1-800-672-5559. NRCS's Internet address is http://www.ncg.nrcs.usda.gov.

CHAPTER 6. PRESENTATION OF EVALUATION RESULTS

6.1 INTRODUCTION

The third, fourth, and fifth chapters of this guidance presented techniques for the collection and analysis of information. Data analysis and interpretation are addressed in detail in Chapter 4 of EPA's *Monitoring Guidance for Determining the Effectiveness of Nonpoint Source Controls* (USEPA, 1997). This chapter provides ideas for the presentation of results.

The presentation of MM or BMP implementation survey results, whether written or oral, is an integral part of a successful monitoring study. A presentation conveys important information from the implementation survey to those who need it (e.g., managers, the public). Failure to present the information in a usable, understandable form results in the data collection effort being an end in itself, and the implementation survey itself might then be considered a failure.

The technical quality of the presentation of results is dependent on at least four criteria—it must be complete, accurate, clear, and concise (Churchill, 1983). Completeness means that the presentation provides all necessary information to the audience in the language that it understands; accuracy is determined by how well an investigator handles the data, phrases findings, and reasons; clarity is the result of clear and logical thinking and a precision of expression; and conciseness is the result of selecting for inclusion only that which is necessary.

Throughout the process of preparing the results of a MM or BMP implementation survey for presentation, it must be kept in mind that the study was initially undertaken to provide information for management purposes—specifically, to help make a decision (Tull and Hawkins, 1990). The presentation of results should be built around the information that was to be developed and the decisions to be made. The message of the presentation must also be tailored to that decision. It must be realized that there will be a time lag between the implementation survey and the presentation of the results, and the results should be presented in light of their applicability to the management decision to be made based on them. The length of the time lag is a key factor in determining this applicability. If the time lag is significant, it should be made clear during the presentation that the situation might have changed since the survey was conducted. If reliable trend data are available, the person making the presentation might be able to provide a sense of the likely magnitude of any change in the situation. If the change in status is thought to be insignificant, evidence should be presented to support this claim. For example, state that "At the time that the implementation survey was conducted, homeowners were using BMPs with increasing frequency, and the lack of any changes in program implementation coupled with continued interaction with homeowners provides no reason to believe that this trend has changed since that time." It would be misleading to state "The monitoring study indicates that homeowners *are* using BMPs with increasing frequency." The validity and force of the message will be enhanced further through use of the active voice (*we believe*) rather than the passive voice (*it is believed*).

Three major factors must be considered when presenting the results of MM and BMP implementation studies: Identifying the target audience, selecting the appropriate medium (printed word, speech, pictures, etc.), and selecting the most appropriate format to meet the needs of the audience.

6.2 Audience Identification

Identification of the audience(s) to which the results of the MM and BMP implementation survey will be presented determines the content and format of the presentation. For results of implementation survey studies, there are typically seven potential audiences:

- Interested/concerned citizens
- Developers/landowners
- Media/general public
- Policy makers
- Resource managers
- Scientists
- School groups

These audiences have different information needs, interests, and abilities to understand complex data. It is the job of the person(s) preparing the presentation to analyze these factors prior to preparing a presentation. The four criteria for presentation quality apply regardless of the audience. Other elements of a comprehensive presentation, such as discussion of the objectives and limitations of the study and necessary details of the method, must be part of the presentation and must be tailored to the audience. For instance, details of the sampling plan, why the plan was chosen over others, and the statistical methods used for analysis might be of interest to other investigators planning a similar study, and such details should be recorded even if they are not part of any presentation of results because of their value for future reference when the monitoring is repeated or similar studies are undertaken, but they are best not included in a presentation to management.

6.3 Presentation Format

Regardless of whether the results of a implementation survey are presented written or orally, or both, the information being presented must be understandable to the audience. Consideration of who the audience is will help ensure that the presentation is particularly suited to its needs, and choice of the correct format for the presentation will ensure that the information is conveyed in a manner that is easy to comprehend.

Most reports will have to be presented both written and orally. Written reports are valuable for peer review, public information dissemination, and for future reference. Oral presentations are often necessary for managers, who usually do not have time to read an entire report, only have need for the results of the study, and are usually not interested in the finer details of the study. Different versions of a report might well have to be written—for the public, scientists, and managers (i.e., an executive summary)—and separate oral presentations for different audiences—the public, developers, managers, and scientists at a conference—might have to be prepared.

Most information can most effectively be presented in the form of tables, charts, and diagrams (Tull and Hawkins, 1990). These graphic forms of data and information presentation can help simplify the presentation, making it easier for an audience to comprehend than if explained exhaustively with words. Words are important for pointing out significant ideas or findings, and for interpreting the results where appropriate. Words should not be used to repeat what is already adequately explained in graphics, and slides or transparencies that are composed largely of words should contain only a few essential ideas each. Presentation of too much written information on a single slide or transparency only confuses the audience. Written slides or transparencies should also be free of graphics, such as clever logos or background highlights—unless the pictures are essential to understanding the information presented—since they only make the slides or transparencies more difficult to read. Examples of graphics and written slides are presented in Figures 6-1 through 6-4.

Different types of graphics have different uses as well. Information presented in a tabular format can be difficult to interpret because the reader has to spend some time with the information to extract the essential points from it. The same information presented in a pie chart or bar graph can convey essential information immediately and avoid the inclusion of background data that are not essential to the point. When preparing

5 Leading Sources of Water Quality Impairment
in various types of water bodies

RANK	ESTUARIES	LAKES	RIVERS
1	***Urban Runoff***	Agriculture	Agriculture
2	STPs	STPs	STPs
3	Agriculture	***Urban Runoff***	Habitat Modification
4	Industry Point Sources	Other NPS	***Urban Runoff***
5	Petroleum Activities	Habitat Modification	Resource Extraction

Figure 6-1. Example of presentation of information in a written slide. (Source: USEPA, 1995)

EROSION AND SEDIMENT CONTROLS

- Sediment loading rates from construction sites are 5-500 times greater than from undeveloped land
- Structural ESC controls can reduce sediment loadings 40-99%
- Structural ESC controls are REQUIRED on all construction sites

Figure 6-2. Example written presentation slide.

information for a report, an investigator should organize the information in various ways and choose that which conveys only the information essential for the audience in the least complicated manner.

6.3.1 *Written Presentations*

The following criteria should be considered when preparing written material:

- *Reading level or level of education* of the target audience.
- *Level of detail necessary* to make the results understandable to the target audience)different audiences require various levels of background information to fully understand the study's results.
- *Layout*. The integration of text, graphics, color, white space, columns, sidebars, and other design elements is important in the production of material that the target audience will find readable and visually appealing.
- *Graphics*. Photos, drawings, charts, tables, maps, and other graphic elements can be used to effectively present information that the reader might otherwise not understand.

6.3.2 *Oral Presentations*

An effective oral presentation requires special preparation. Tull and Hawkins (1990) recommend three steps:

Leading Sources of Pollution
Relative Quantity of Lake Acres Affected by Source
Urban Runoff
Municipal Point Sources
Habitat Modification
Agriculture
Other Sources

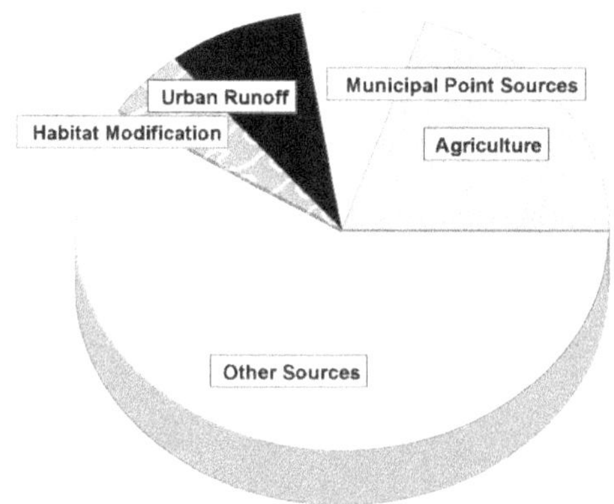

Figure 6-3. Example representation of data in the form of a pie chart

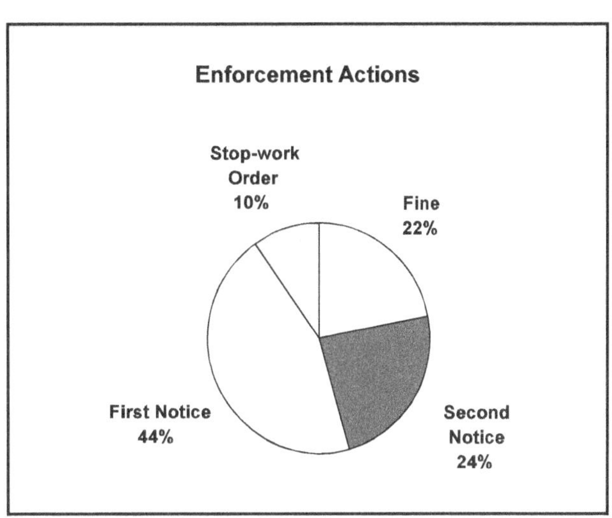

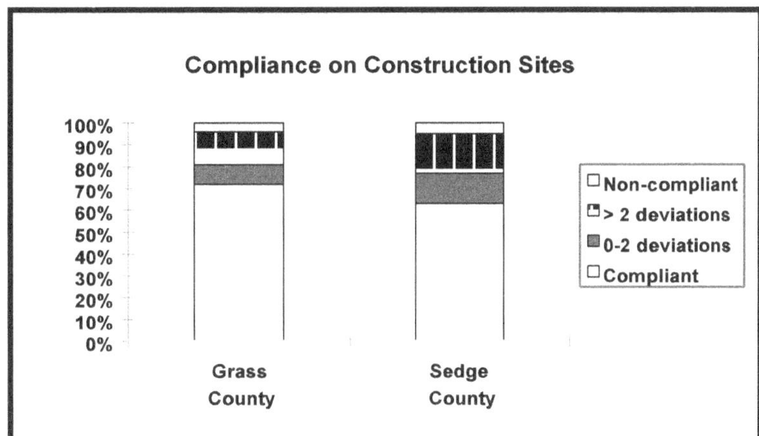

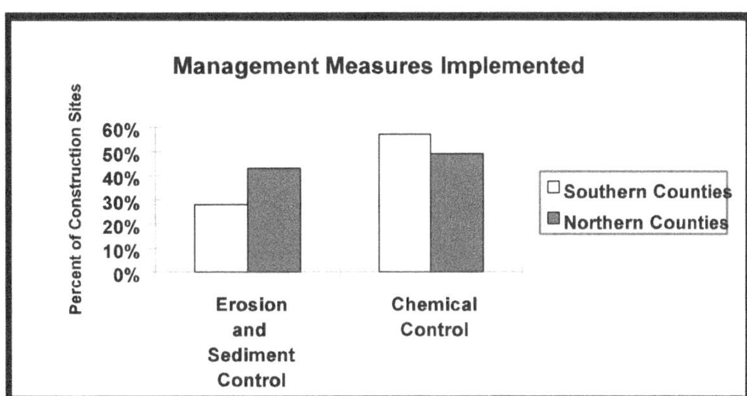

Figure 6-4. Graphical representations of data from construction site surveys.

1. Analyze the audience, as explained above;
2. Prepare an outline of the presentation, and preferably a written script;
3. Rehearse it. Several dry runs of the presentation should be made, and if possible it should be taped on a VCR and the presentation analyzed.

These steps are extremely important if an oral presentation is to be effective. Remember that oral presentations of ½ to 1 hour are often all that is available for the presentation of the results of months of research to managers who are poised to make decisions based on the presentation. Adequate preparation is essential if the oral presentation is to accomplish its purpose.

6.4 For Further Information

The provision of specific examples of effective and ineffective presentation graphics, writing styles, and organizations is beyond the scope of this document. A number of resources that contain suggestions for how study results should be presented are available, however, and should be consulted. A listing of some references is provided below.

- *The New York Public Library Writer's Guide to Style and Usage* (NYPL, 1994) has information on design, layout, and presentation in addition to guidance on grammar and style.
- *Good Style: Writing for Science and Technology* (Kirkman, 1992) provides techniques for presenting technical material in a coherent, readable style.
- *The Modern Researcher* (Barzun and Graff, 1992) explains how to turn research into readable, well organized writing.
- *Writing with Precision: How to Write So That You Cannot Possibly Be Misunderstood*, 6th ed. (Bates, 1993) addresses communication problems of the 1990s.
- *Designer's Guide to Creating Charts & Diagrams* (Holmes, 1991) gives tips for combining graphics with statistical information.
- *The Elements of Graph Design* (Kosslyn, 1993) shows how to create effective displays of quantitative data.

REFERENCES

Academic Press. 1992. *Dictionary of Science and Technology*. Academic Press, Inc., San Diego, CA.

Adams, T. 1994. *Implementation monitoring of forestry best management practices on harvested sites in South Carolina*. Best Management Practices Monitoring Report BMP-2. South Carolina Forestry Commission, Columbia, South Carolina. September.

Barbé, D.E., H. Miller, and S. Jalla. 1993. Development of a computer interface among GDS, SCADA, and SWMM for use in urban runoff simulation. In *Symposium on Geographic Information Systems and Water Resources*, American Water Resources Association, Mobile, Alabama, March 14-17, pp. 113-120.

Barzun, J., and H.F. Graff. 1992. *The Modern Researcher.* 5th ed. Houghton Mifflin.

Bates, J. 1993. *Writing with Precision: How to Write So That You Cannot Possibly Be Misunderstood.* 6th ed. Acropolis.

Blalock, H.M., Jr. 1979. *Social Statistics.* Rev. 2nd ed. McGraw-Hill Book Company, New York, NY.

BLM. 1991. *Inventory and Monitoring Coordination: Guidelines for the Use of Aerial Photography in Monitoring.* Technical Report TR 1734-1. Department of the Interior, Bureau of Land Management, Reston, VA.

Born, J.D., and D.D. Van Hooser. 1988. Intermountain Research Station remote sensing use for resource inventory, planning, and monitoring. In *Remote Sensing for Resource Inventory, Planning, and Monitoring.* Proceedings of the Second Forest Service Remote Sensing Applications Conference, Sidell, Louisiana, and NSTL, Mississippi, April 11-15, 1988.

Casley, D.J., and D.A. Lury. 1982. *Monitoring and Evaluation of Agriculture and Rural Development Projects.* The Johns Hopkins University Press, Baltimore, MD.

Center for Watershed Protection (CWP). 1997. Delaware program improves construction site inspection. Technical Note No. 85. Center for Watershed Protection, Silver Spring, Maryland. *Watershed Protection Techniques* 2(3):440-442.

Churchill, G.A., Jr. 1983. *Marketing Research: Methodological Foundations,* 3rd ed. The Dryden Press, New York, NY.

Clayton and Brown. 1996. *Environmental indicators to assess stormwater control programs and practices.* Center for Watershed Protection, Silver Sprint, Maryland.

Cochran, W.G. 1977. *Sampling techniques.* 3rd ed. John Wiley and Sons, New York, New York.

Conover, W.J. 1980. *Practical Nonparametric Statistics*, 2nd ed. Wiley, New York.

Cooper, B., and S. Carson. 1993. Application of a geographic information system (GIS) to groundwater assessment: A case study in Loudoun County, Virginia. In *Symposium on Geographic Information Systems and Water Resources*, American Water Resources Association, Mobile, Alabama, March 14-17, pp. 331-341.

Cross-Smiecinski, A., and L.D. Stetzenback. 1994. *Quality planning for the life science researcher: Meeting quality assurance requirements*. CRC Press, Boca Raton, Florida.

CTIC. 1994. *1994 National Crop Residue Management Survey.* Conservation Technology Information Center, West Lafayette, IN.

CTIC. 1995. *Conservation IMPACT,* vol. 13, no. 4, April 1995. Conservation Technology Information Center, West Lafayette, IN.

Delaware DNREC. 1996. *COMPAS Delaware: An integrated nonpoint source pollution information system.* Delaware Department of Natural Resources and Environmental Control, Dover, Delaware. April.

Environmental Law Institute. 1997. *Enforceable State Mechanisms for the Control of Nonpoint Source Water Pollution.* Environmental Law Institute Project #970300. Washington, DC.

Ferber, R., D.F. Blankertz, and S. Hollander. 1964. *Marketing Research.* The Ronald Press Company, New York, NY.

Freund, J.E. 1973. *Modern elementary statistics.* Prentice-Hall, Englewood Cliffs, New Jersey.

Galli, J., and L. Herson. 1989. *Anacostia River Basin stormwater retrofit inventory, 1989, Prince George's County*. Prince George's County Department of Environmental Resources.

Gaugush, R.F. 1987. *Sampling Design for Reservoir Water Quality Investigations.* Instruction Report E-87-1. Department of the Army, US Army Corps of Engineers, Washington, DC.

Gene Kroupa & Associates. 1995. *Westmorland lawn care survey.* Prepared for Wisconsin Department of Natural Resources, Division of Water Resources Management. April.

Gilbert, R.O. 1987. *Statistical Methods for Environmental Pollution Monitoring*. Van Nostrand Reinhold, New York, NY.

Hackett, R.L. 1988. Remote sensing at the North Central Forest Experiment Station. In *Remote Sensing for Resource Inventory, Planning, and Monitoring.* Proceedings of the Second Forest Service Remote Sensing Applications Conference, Sidell, Louisiana, and NSTL, Mississippi, April 11-15, 1988.

Hall, R.J., and A.H. Aldred. 1992. Forest regeneration appraisal with large-scale aerial photographs. *The Forestry Chronicle* 68(1):142-150.

Helsel, D.R., and R.M. Hirsch. 1995. *Statistical Methods in Water Resources.* Elsevier. Amsterdam.

Hetzel, G.E. 1988. Remote sensing applications and monitoring in the Rocky Mountain region. In *Remote Sensing for Resource Inventory, Planning, and Monitoring.* Proceedings of the Second Forest Service Remote Sensing Applications Conference, Sidell, Louisiana, and NSTL, Mississippi, April 11-15, 1988.

Holmes, N. 1991. *Designer's Guide to Creating Charts & Diagrams.* Watson-Guptill.

Hook, D., W. McKee, T. Williams, B. Baker, L. Lundquist, R. Martin, and J. Mills. 1991. *A Survey of Voluntary Compliance of Forestry BMPs.* South Carolina Forestry Commission, Columbia, SC.

Hudson, W.D. 1988. Monitoring the long-term effects of silvicultural activities with aerial photography. *J. Forestry* (March):21-26.

IDDHW. 1993. *Forest Practices Water Quality Audit 1992.* Idaho Department of Health and Welfare, Division of Environmental Quality, Boise, ID.

Kirkman, J. 1992. *Good Style: Writing for Science and Technology.* Chapman and Hall.

Kosslyn, S.M. 1993. *The Elements of Graph Design.* W.H. Freeman.

Kroll, R., and D.L. Murphy. 1994. *Residential pesticide usage survey.* Technical Report No. 94-011. Maryland Department of the Environment, Water Management Administration, Water Quality Program.

Kupper, L.L., and K.B. Hafner. 1989. How appropriate are popular sample size formulas? *Amer. Statistician* 43:101-105.

Lindsey, G., L. Roberts, and W. Page. 1992. Maintenance of stormwater BMPs in four Maryland counties: A status report. *J. Soil Water Conserv.* 47(5):417-422.

MacDonald, L.H., A.W. Smart, and R.C. Wissmar. 1991. *Monitoring guidelines to evaluate the effects of forestry activities on streams in the pacific northwest and Alaska.* EPA/910/9-91-001. U.S. Environmental Protection Agency, Region 10, Seattle, Washington.

Mann, H.B., and D.R. Whitney. 1947. On a test of whether one of two random variables is stochastically larger than the other. *Annals of Mathematical Statistics* 18:50-60.

McNew, R.W. 1990. Sampling and Estimating Compliance with BMPs. In *Workshop on Implementation Monitoring of Forestry Best Management Practices*, Southern Group of State Foresters, USDA Forest Service, Southern Region, Atlanta, GA, January 23-25, 1990, pp. 86-105.

Meals, D.W. 1988. *Laplatte River Watershed Water Quality Monitoring & Analysis Program.* Program Report No. 10. Vermont Water Resources Research Center, School of Natural Resources, University of Vermont, Burlington, VT.

Mereszczak, I. 1988. Applications of large format camera—color infrared photography to monitoring vegetation management within the scope of forest plans. In *Remote Sensing for Resource Inventory, Planning, and Monitoring.* Proceedings of the Second Forest Service Remote Sensing Applications Conference, Sidell, Louisiana, and NSTL, Mississippi, April 11-15, 1988.

NYPL. 1994. *The New York Public Library Writer's Guide to Style and Usage.* A Stonesong Press book. Harper Collins Publishers, New York, NY.

Ohrel, R.L. 1996. *Technical memorandum: Survey of local erosion and sediment control programs.* Center for Watershed Protection, Silver Spring, Maryland.

Owens, T. 1988. Using 35mm photographs in resource inventories. In *Remote Sensing for Resource Inventory, Planning, and Monitoring.* Proceedings of the Second Forest Service Remote Sensing Applications Conference, Sidell, Louisiana, and NSTL, Mississippi, April 11-15, 1988.

Paterson, R.G. 1994. Construction practices: The good, the bad, and the ugly. *Watershed Protection Techniques 1(3):95-99.*

Pelletier, R.E., and R.H. Griffin. 1988. An evaluation of photographic scale in aerial photography for identification of conservation practices. *J. Soil Water Conserv.* 43(4):333-337.

Pensyl, L.K., and P.F. Clement. 1987. *Results of the state of Maryland infiltration practices survey.* Presented at the state of Maryland sediment and stormwater management conference, Washington College, Chestertown, Maryland.

Rashin, E., C. Clishe, and A. Loch. 1994. *Effectiveness of forest road and timber harvest best management practices with respect to sediment-related water quality impacts.* Interim Report No. 2. Washington State Department of Ecology, Environmental Investigations and Laboratory Services program, Wastershed Assessments Section. Ecology Publication No. 94-67. Olympia, Washington.

Remington, R.D., and M.A. Schork. 1970. *Statistics with applications to the biological and health sciences.* Prentice-Hall, Englewood Cliffs, New Jersey.

Reutebuch, S.E. 1988. A method to control large-scale photos when surveyed ground control is unavailable. In *Remote Sensing for Resource Inventory, Planning, and Monitoring.* Proceedings of the Second Forest Service Remote Sensing Applications Conference, Sidell, Louisiana, and NSTL, Mississippi, April 11-15, 1988.

Richards, P.A. 1993. Louisiana discharger inventory geographic information system. In *Symposium on Geographic Information Systems and Water Resources,* American Water Resources Association, Mobile, Alabama, March 14-17, pp. 507-516.

Robinson, K.J., and R.M. Ragan. 1993. Geographic information system based nonpoint pollution modeling. In *Symposium on Geographic Information Systems and Water Resources,* American Water Resources Association, Mobile, Alabama, March 14-17, pp. 53-60.

Rossman, R., and M.J. Phillips. 1991. *Minnesota forestry best management practices implementation monitoring. 1991 forestry field audit.* Minnesota Department of Natural Resources, Division of Forestry.

Sanders, T.G., R.C. Ward, J.C. Loftis, T.D. Steele, D.D. Adrian, and V. Yevjevich. 1983. *Design Networks for Monitoring Water Quality.* Water Resources Publications, Littleton, CO.

Schultz, B. 1992. *Montana Forestry Best Management Practices Implementation Monitoring.* The 1992 Forestry BMP Audits Final Report. Montana Department of State Lands, Forestry Division, Missoula, MT.

Snedecor, G.W. and W.G. Cochran. 1980. *Statistical methods.* 7th ed. The Iowa State University Press, Ames, Iowa.

Tull, D.S., and D.I. Hawkins. 1990. *Marketing Research: Measurement and Method.* 5th ed. Macmillan Publishing Company, New York, NY.

US Census Bureau. 1990. *1990 Census of the United States, Census of Housing, Sewage Disposal.* U.S. Bureau of the Census, Washington, DC.

USDA. 1994. *1992 National Resources Inventory.* U.S. Department of Agriculture, Natural Resource Conservation Service, Resources Inventory and Geographical Information Systems Division, Washington, DC.

USEPA. 1993. *Water Quality Effects And Nonpoint Source Pollution Control For Forestry: An Annotated Bibliography.* U.S. Environmental Protection Agency, Office of Water, Washington, DC.

USEPA. 1993a. *Evaluation of the Experimental Rural Clean Water Program.* EPA 841-R-93-005. U.S. Environmental Protection Agency, Office of Water, Washington, DC.

USEPA. 1993b. *Guidance Specifying Management Measures For Sources Of Nonpoint Pollution In Coastal Waters.* EPA 840-B-92-002. U.S. Environmental Protection Agency, Office of Water, Washington, DC.

USEPA. 1994. *National Water Quality Inventory: 1992 Report to Congress.* EPA 841-R-94-001. U.S. Environmental Protection Agency, Office of Water, Washington, DC.

USEPA. 1995. *National water quality inventory 1994 Report to Congress.* EPA 841-R-95-005. U.S. Environmental Protection Agency, Office of Water, Washington, DC.

USEPA. 1997. *Monitoring guidance for determining the effectiveness of nonpoint source controls.* EPA841-B-96-004. U.S. Environmental Protection Agency, Office of Water, Washington, DC. September.

USEPA. 1997a. *Management of Onsite Wastewater Systems.* Draft. United States Environmental Protection Agency, Office of Wetlands, Oceans, and Watersheds. Washington, DC.

USEPA. (undated). *Buzzards Bay "SepTrack" initiative*. U.S. Environmental Protection Agency, Washington, DC.

USGS. 1990. *Land Use and Land Cover Digital Data from 1:250,000- and 1:100,000-Scale Maps: Data Users Guide.* National Mapping Program Technical Instructions Data Users Guide 4. U.S. Department of the Interior, U.S. Geological Survey, Reston, VA.

WADOE. 1994. *Effectiveness of Forest Road and Timber Harvest Best Management Practices With Respect to Sediment-Related Water Quality Impacts.* Washington State Department of Ecology, Environmental Investigations and Laboratory Services Program, Watershed Assessments Section. Ecology Publication No. 94-67. Olympia, WA.

Washington State. 1996. *Guidance handbook for onsite sewage system monitoring programs in Washington State*. Washington State Department of Health, Community Environmental Health Programs. Olympia, Washington. Cited in USEPA, 1997a.

Weston, R.F. 1979. *Management of onsite and alternative wastewater systems*. Draft. Prepared for U.S. Environmental Protection Agency, Cincinnati, Ohio. Cited in USEPA, 1997a.

Wilcoxon, F. 1945. Individual comparisons by ranking metheds. *Biometrics* 1:80-83.

Winer, B.J. 1971. *Statistical principles in experimental design.* McGraw-Hill Book Company, New York.

GLOSSARY

accuracy: the extent to which a measurement approaches the true value of the measured quantity.

aerial photography: the practice of taking photographs from an airplane, helicopter, or other aviation device while it is airborne.

allocation, Neyman: stratified sampling in which the cost of sampling each stratum is in proportion to the size of the stratum but variability between strata changes.

allocation, proportional: stratified sampling in which the variability and cost of sampling for each stratum are in proportion to the size of the stratum.

allowable error: the level of error acceptable for the purposes of a study.

ANOVA: see test, analysis of variance.

assumptions: characteristics of a population of a sampling method taken to be true without proof.

bar graph: a representation of data wherein data are grouped and represented as vertical or horizontal bars over an axis.

best professional judgment: an informed opinion made by a professional in the appropriate field of study or expertise.

best management practice: a practice or combination of practices that are determined to be the most effective and practicable means of controlling point and/or nonpoint pollutants at levels compatible with environmental quality goals.

bias: a characteristic of samples such that when taken from a population with a known parameter, their average does not give the parametric value.

binomial: an algebraic expression that is the sum or difference of two terms.

camera format: refers to the size of the negative taken by a camera. 35mm is a small camera format.

chi-square distribution: a scaled quantity whose distribution provides the distribution of the sample variance.

coefficient of variation: a statistical measure used to compare the relative amounts of variation in populations having different means.

confidence interval: a range of values about a measured value in which the true value is presumed to lie.

consistency: conforming to a regular method or style; an approach that keeps all factors of measurement similar from one measurement to the next to the extent possible.

cumulative effects: the total influences attributable to numerous individual influences.

degrees of freedom: the number of residuals (the difference between a measured value and the sample average) required to completely determine the others.

design, balanced: a sampling design wherein separate sets of data to be used are similar in quantity and type.

distribution: the allocation or spread of values of a given parameter among its possible values.

erosion potential: a measure of the ease with which soil can be carried away in storm water runoff or irrigation runoff.

error: the fluctuation that occurs from one repetition to another; also *experimental error*.

estimate, baseline: an estimate of baseline, or actual conditions.

estimate, pooled: a single estimate obtained from combining several individual estimates to obtain a single value.

finite population correction term: a correction term used when population size is small relative to sample size.

hydrologic modification: the alteration of the natural circulation or distribution of water by the placement of structures or other activities.

hypothesis, alternative: the hypothesis that is contrary to the null hypothesis.

hypothesis, null: the hypothesis or conclusion assumed to be true prior to any analysis.

management measure: an economically achievable measure for the control of the addition of pollutants from existing and new categories and classes of nonpoint sources of pollution, which reflect the greatest degree of pollutant reduction achievable through the application of the best available nonpoint pollution control practices, technologies, processes, siting criteria, operating methods, or other alternatives.

mean, estimated: a value of population mean arrived at through sampling.

mean, overall: the measured average of a population.

mean, stratum: the measured average within a sample subgroup or stratum.

measurement bias: a consistent under- or overestimation of the true value of something being measured, often due to the method of measurement.

measurement error: the deviation of a measurement from the true value of that which is being measured.

median: the value of the middle term when data are arranged in order of size; a measure of central tendency.

monitoring, baseline: monitoring conducted to establish initial knowledge about the actual state of a population.

monitoring, compliance: monitoring conducted to determine whether those who must implement programs, best management practices, or management measures, or who must conduct operations according to standards or specifications, are doing so.

monitoring, project: monitoring conducted to determine the impact of a project, activity, or program.

monitoring, validation: monitoring conducted to determine how well a model accurately reflects reality.

navigational error: error in determining the actual location (altitude or latitude/longitude) of an airplane or other aviation device due to instrumentation or the operator.

nominal: referred to by name; variables that cannot be measured but must be expressed qualitatively.

nonparametric method: distribution-free method; any of various inferential procedures whose conclusions do not rely on assumptions about the distribution of the population of interest.

normal approximation: an assumption that a population has the characteristics of a normally distributed population.

normal deviate: deviation from the mean expressed in units of σ.

ordinal: ordered such that the position of an element in a series is specified.

parametric method: any statistical method whose conclusions rely on assumptions about the distribution of the population of interest.

physiography: a description of the surface features of the earth; a description of landforms.

pie chart: a representation of data wherein data are grouped and represented as more or less triangular sections of a circle and the total is the entire circle.

population, sample: the members of a population that are actually sampled or measured.

population, target: the population about which inferences are made; the group of interest, from which samples are taken.

population unit: an individual member of a target population that can be measured independently of other members.

power: the probability of correctly rejecting the null hypothesis when the alternative hypothesis is false.

precision: a measure of the similarity of individual measurements of the same population.

question, dichotomous: a question that allows for only two responses, such as "yes" and "no".

question, double-barreled: two questions asked as a single question.

question, multiple-choice: a question with two or more predetermined responses.

question, open-ended: a question format that requires a response beyond "yes" or "no".

remote sensing: methods of obtaining data from a location distant from the object being measured, such as from an airplane or satellite.

resolution: the sharpness of a photograph.

sample size: the number of population units measured.

sampling, cluster: sampling in which small groups of population units are selected for sampling and each unit in each selected group is measured.

sampling, simple random: sampling in which each unit of the target population has an equal chance of being selected.

sampling, stratified random: sampling in which the target population is divided into separate subgroups, each of which is more internally similar than the overall population is, prior to sample selection.

sampling, systematic: sampling in which population units are chosen in accordance with a predetermined sample selection system.

sampling error: error attributable to actual variability in population units not accounted for by the sampling method.

scale (aerial photography): the proportion of the image size of an object (such as a land area) to its actual size, e.g., 1:3000. The smaller the second number, the larger the scale.

scale system: a system for ranking measurements or members of a population on a scale, such as 1 to 5.

significance level: in hypothesis testing, the probability of rejecting a hypothesis that is correct, that is, the probability of a Type I error.

standard deviation: a measure of spread; the positive square root of the variance.

standard error: an estimate of the standard deviation of means that would be expected if a collection of means based on equal-sized samples of n items from the same population were obtained.

statistical inference: conclusions drawn about a population using statistics.

statistics, descriptive: measurements of population characteristics designed to summarize important features of a data set.

stratification: the process of dividing a population into internally similar subgroups.

stratum: one of the subgroups created prior to sampling in stratified random sampling.

subjectivity: a characteristic of analysis that requires personal judgement on the part of the person doing the analysis.

target audience: the population that a monitoring effort is intended to measure.

test, analysis of variance (ANOVA): a statistical test used to determine whether two or more sample means could have been obtained from populations with the same parametric mean.

test, Friedman: a nonparametric test that can be used for analysis when two variables are involved.

test, Kruskal-Wallis: a nonparametric test recommended for the general case with a samples and n_i variates per sample.

test, Mann-Whitney: a nonparametric test for use when a test is only between two samples.

test, Student's t: a statistical test used to test for significant differences between means when only two samples are involved.

test, Tukey's: a test to ascertain whether the interaction found in a given set of data can be explained in terms of multiplicative main effects.

test, Wilcoxon's: a nonparametric test for use when only two samples are involved.

total maximum daily load: a total allowable addition of pollutants from all affecting sources to an individual waterbody over a 24-hour period.

transformation, data: manipulation of data such that they will meet the assumptions required for analysis.

unit sampling cost: the cost attributable to sampling a single population unit.

variance: a measure of the spread of data around the mean.

watershed assessment: an investigation of numerous characteristics of a watershed in order to describe its actual condition.

INDEX

APPENDIX A

Statistical Tables

Table A1. Cumulative areas under the Normal distribution (values of p corr to Z_p)

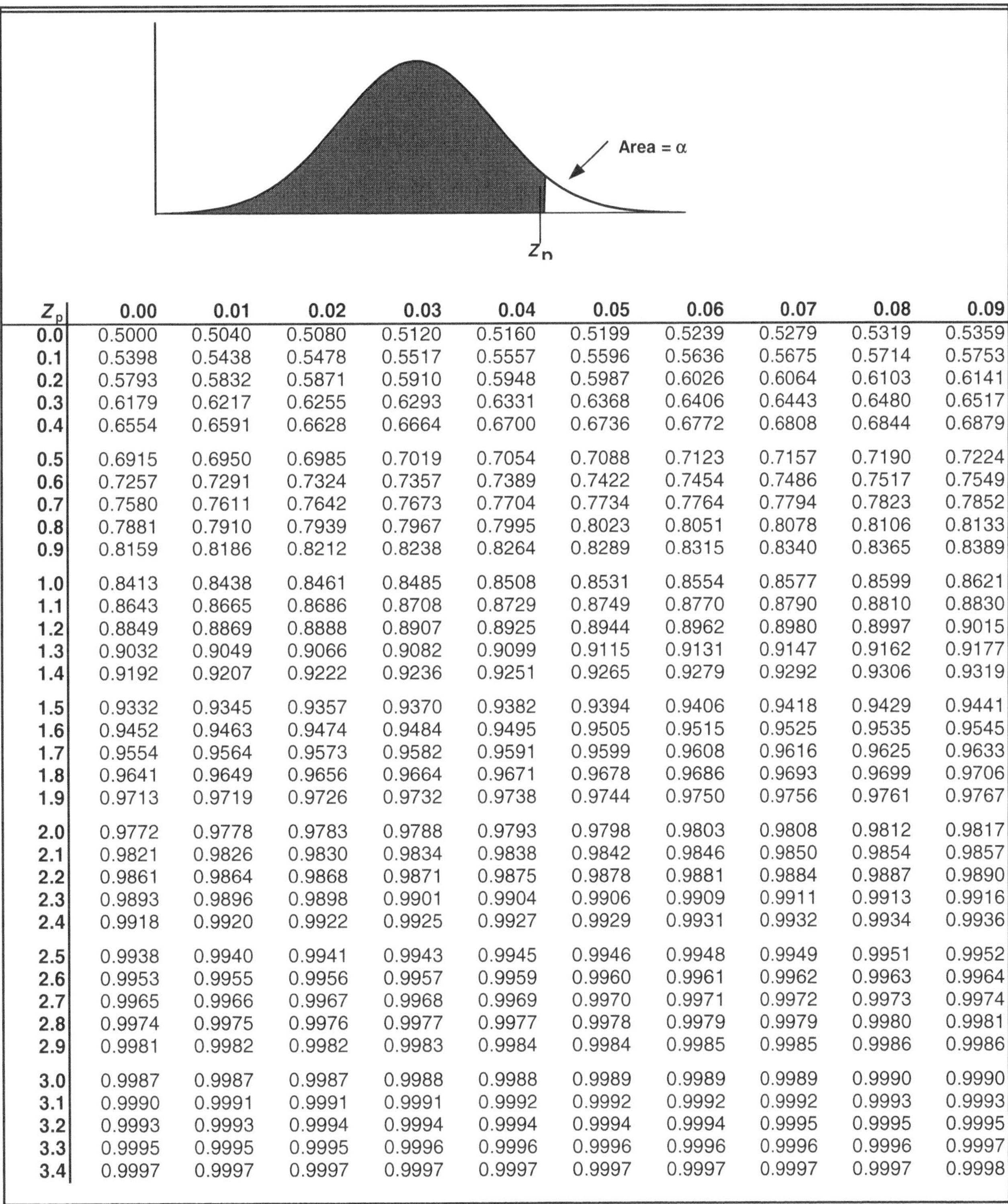

Z_p	0.00	0.01	0.02	0.03	0.04	0.05	0.06	0.07	0.08	0.09
0.0	0.5000	0.5040	0.5080	0.5120	0.5160	0.5199	0.5239	0.5279	0.5319	0.5359
0.1	0.5398	0.5438	0.5478	0.5517	0.5557	0.5596	0.5636	0.5675	0.5714	0.5753
0.2	0.5793	0.5832	0.5871	0.5910	0.5948	0.5987	0.6026	0.6064	0.6103	0.6141
0.3	0.6179	0.6217	0.6255	0.6293	0.6331	0.6368	0.6406	0.6443	0.6480	0.6517
0.4	0.6554	0.6591	0.6628	0.6664	0.6700	0.6736	0.6772	0.6808	0.6844	0.6879
0.5	0.6915	0.6950	0.6985	0.7019	0.7054	0.7088	0.7123	0.7157	0.7190	0.7224
0.6	0.7257	0.7291	0.7324	0.7357	0.7389	0.7422	0.7454	0.7486	0.7517	0.7549
0.7	0.7580	0.7611	0.7642	0.7673	0.7704	0.7734	0.7764	0.7794	0.7823	0.7852
0.8	0.7881	0.7910	0.7939	0.7967	0.7995	0.8023	0.8051	0.8078	0.8106	0.8133
0.9	0.8159	0.8186	0.8212	0.8238	0.8264	0.8289	0.8315	0.8340	0.8365	0.8389
1.0	0.8413	0.8438	0.8461	0.8485	0.8508	0.8531	0.8554	0.8577	0.8599	0.8621
1.1	0.8643	0.8665	0.8686	0.8708	0.8729	0.8749	0.8770	0.8790	0.8810	0.8830
1.2	0.8849	0.8869	0.8888	0.8907	0.8925	0.8944	0.8962	0.8980	0.8997	0.9015
1.3	0.9032	0.9049	0.9066	0.9082	0.9099	0.9115	0.9131	0.9147	0.9162	0.9177
1.4	0.9192	0.9207	0.9222	0.9236	0.9251	0.9265	0.9279	0.9292	0.9306	0.9319
1.5	0.9332	0.9345	0.9357	0.9370	0.9382	0.9394	0.9406	0.9418	0.9429	0.9441
1.6	0.9452	0.9463	0.9474	0.9484	0.9495	0.9505	0.9515	0.9525	0.9535	0.9545
1.7	0.9554	0.9564	0.9573	0.9582	0.9591	0.9599	0.9608	0.9616	0.9625	0.9633
1.8	0.9641	0.9649	0.9656	0.9664	0.9671	0.9678	0.9686	0.9693	0.9699	0.9706
1.9	0.9713	0.9719	0.9726	0.9732	0.9738	0.9744	0.9750	0.9756	0.9761	0.9767
2.0	0.9772	0.9778	0.9783	0.9788	0.9793	0.9798	0.9803	0.9808	0.9812	0.9817
2.1	0.9821	0.9826	0.9830	0.9834	0.9838	0.9842	0.9846	0.9850	0.9854	0.9857
2.2	0.9861	0.9864	0.9868	0.9871	0.9875	0.9878	0.9881	0.9884	0.9887	0.9890
2.3	0.9893	0.9896	0.9898	0.9901	0.9904	0.9906	0.9909	0.9911	0.9913	0.9916
2.4	0.9918	0.9920	0.9922	0.9925	0.9927	0.9929	0.9931	0.9932	0.9934	0.9936
2.5	0.9938	0.9940	0.9941	0.9943	0.9945	0.9946	0.9948	0.9949	0.9951	0.9952
2.6	0.9953	0.9955	0.9956	0.9957	0.9959	0.9960	0.9961	0.9962	0.9963	0.9964
2.7	0.9965	0.9966	0.9967	0.9968	0.9969	0.9970	0.9971	0.9972	0.9973	0.9974
2.8	0.9974	0.9975	0.9976	0.9977	0.9977	0.9978	0.9979	0.9979	0.9980	0.9981
2.9	0.9981	0.9982	0.9982	0.9983	0.9984	0.9984	0.9985	0.9985	0.9986	0.9986
3.0	0.9987	0.9987	0.9987	0.9988	0.9988	0.9989	0.9989	0.9989	0.9990	0.9990
3.1	0.9990	0.9991	0.9991	0.9991	0.9992	0.9992	0.9992	0.9992	0.9993	0.9993
3.2	0.9993	0.9993	0.9994	0.9994	0.9994	0.9994	0.9994	0.9995	0.9995	0.9995
3.3	0.9995	0.9995	0.9995	0.9996	0.9996	0.9996	0.9996	0.9996	0.9996	0.9997
3.4	0.9997	0.9997	0.9997	0.9997	0.9997	0.9997	0.9997	0.9997	0.9997	0.9998

Table A2. Percentiles of the $t_{\alpha,df}$ distribution (values of t such that 100(1-α distribution is less than t)

Area = α

t

df	α = 0.40	α = 0.30	α = 0.20	α = 0.10	α = 0.05	α = 0.025	α = 0.010	α = 0.005
1	0.3249	0.7265	1.3764	3.0777	6.3137	12.7062	31.8210	63.6559
2	0.2887	0.6172	1.0607	1.8856	2.9200	4.3027	6.9645	9.9250
3	0.2767	0.5844	0.9785	1.6377	2.3534	3.1824	4.5407	5.8408
4	0.2707	0.5686	0.9410	1.5332	2.1318	2.7765	3.7469	4.6041
5	0.2672	0.5594	0.9195	1.4759	2.0150	2.5706	3.3649	4.0321
6	0.2648	0.5534	0.9057	1.4398	1.9432	2.4469	3.1427	3.7074
7	0.2632	0.5491	0.8960	1.4149	1.8946	2.3646	2.9979	3.4995
8	0.2619	0.5459	0.8889	1.3968	1.8595	2.3060	2.8965	3.3554
9	0.2610	0.5435	0.8834	1.3830	1.8331	2.2622	2.8214	3.2498
10	0.2602	0.5415	0.8791	1.3722	1.8125	2.2281	2.7638	3.1693
11	0.2596	0.5399	0.8755	1.3634	1.7959	2.2010	2.7181	3.1058
12	0.2590	0.5386	0.8726	1.3562	1.7823	2.1788	2.6810	3.0545
13	0.2586	0.5375	0.8702	1.3502	1.7709	2.1604	2.6503	3.0123
14	0.2582	0.5366	0.8681	1.3450	1.7613	2.1448	2.6245	2.9768
15	0.2579	0.5357	0.8662	1.3406	1.7531	2.1315	2.6025	2.9467
16	0.2576	0.5350	0.8647	1.3368	1.7459	2.1199	2.5835	2.9208
17	0.2573	0.5344	0.8633	1.3334	1.7396	2.1098	2.5669	2.8982
18	0.2571	0.5338	0.8620	1.3304	1.7341	2.1009	2.5524	2.8784
19	0.2569	0.5333	0.8610	1.3277	1.7291	2.0930	2.5395	2.8609
20	0.2567	0.5329	0.8600	1.3253	1.7247	2.0860	2.5280	2.8453
21	0.2566	0.5325	0.8591	1.3232	1.7207	2.0796	2.5176	2.8314
22	0.2564	0.5321	0.8583	1.3212	1.7171	2.0739	2.5083	2.8188
23	0.2563	0.5317	0.8575	1.3195	1.7139	2.0687	2.4999	2.8073
24	0.2562	0.5314	0.8569	1.3178	1.7109	2.0639	2.4922	2.7970
25	0.2561	0.5312	0.8562	1.3163	1.7081	2.0595	2.4851	2.7874
26	0.2560	0.5309	0.8557	1.3150	1.7056	2.0555	2.4786	2.7787
27	0.2559	0.5306	0.8551	1.3137	1.7033	2.0518	2.4727	2.7707
28	0.2558	0.5304	0.8546	1.3125	1.7011	2.0484	2.4671	2.7633
29	0.2557	0.5302	0.8542	1.3114	1.6991	2.0452	2.4620	2.7564
30	0.2556	0.5300	0.8538	1.3104	1.6973	2.0423	2.4573	2.7500
35	0.2553	0.5292	0.8520	1.3062	1.6896	2.0301	2.4377	2.7238
40	0.2550	0.5286	0.8507	1.3031	1.6839	2.0211	2.4233	2.7045
50	0.2547	0.5278	0.8489	1.2987	1.6759	2.0086	2.4033	2.6778
60	0.2545	0.5272	0.8477	1.2958	1.6706	2.0003	2.3901	2.6603
80	0.2542	0.5265	0.8461	1.2922	1.6641	1.9901	2.3739	2.6387
100	0.2540	0.5261	0.8452	1.2901	1.6602	1.9840	2.3642	2.6259
150	0.2538	0.5255	0.8440	1.2872	1.6551	1.9759	2.3515	2.6090
200	0.2537	0.5252	0.8434	1.2858	1.6525	1.9719	2.3451	2.6006
inf.	0.2533	0.5244	0.8416	1.2816	1.6449	1.9600	2.3264	2.5758

Table A3. Upper and lower percentiles of the Chi-square distribution

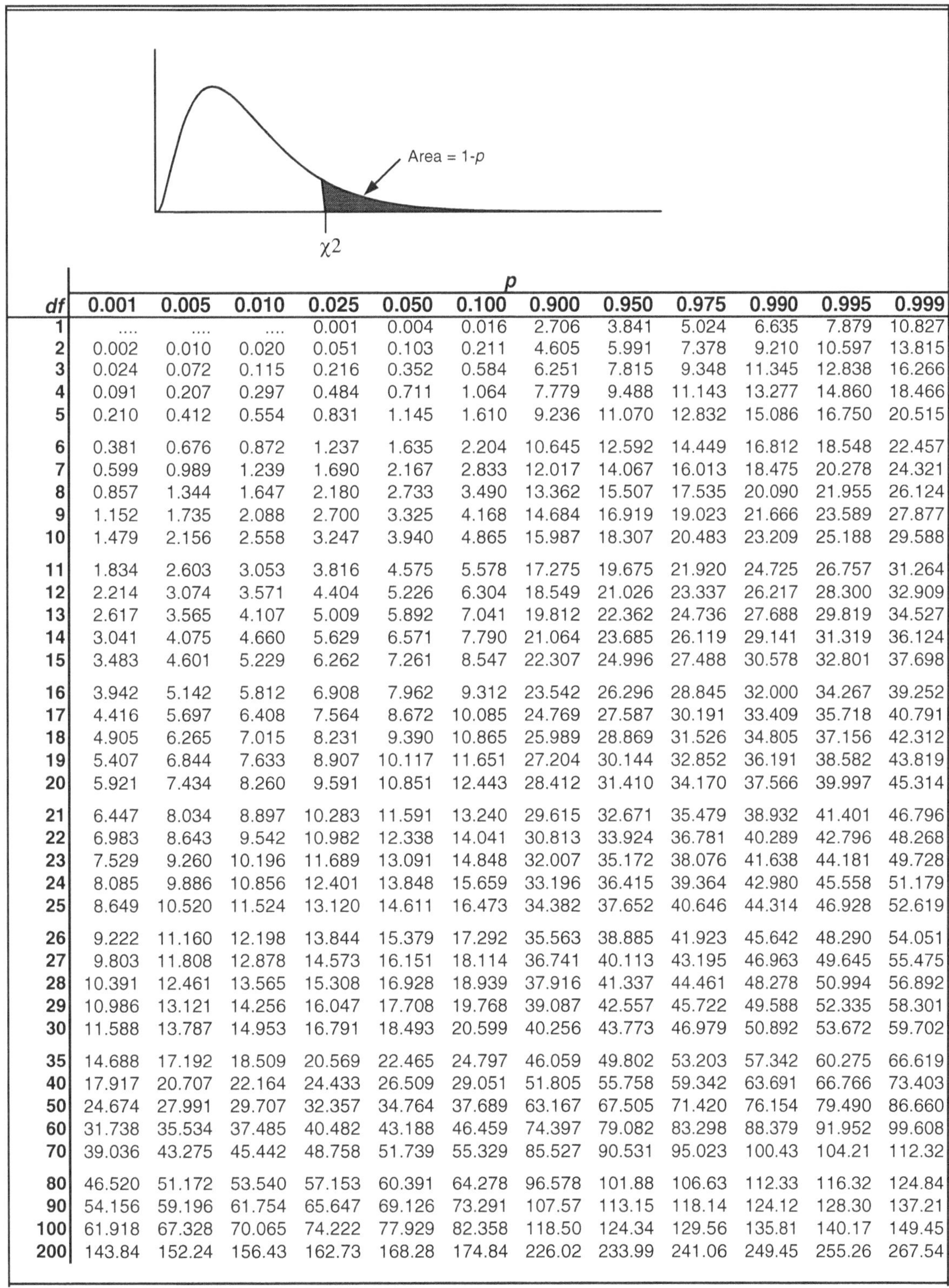

df	p = 0.001	0.005	0.010	0.025	0.050	0.100	0.900	0.950	0.975	0.990	0.995	0.999
1	....	....	....	0.001	0.004	0.016	2.706	3.841	5.024	6.635	7.879	10.827
2	0.002	0.010	0.020	0.051	0.103	0.211	4.605	5.991	7.378	9.210	10.597	13.815
3	0.024	0.072	0.115	0.216	0.352	0.584	6.251	7.815	9.348	11.345	12.838	16.266
4	0.091	0.207	0.297	0.484	0.711	1.064	7.779	9.488	11.143	13.277	14.860	18.466
5	0.210	0.412	0.554	0.831	1.145	1.610	9.236	11.070	12.832	15.086	16.750	20.515
6	0.381	0.676	0.872	1.237	1.635	2.204	10.645	12.592	14.449	16.812	18.548	22.457
7	0.599	0.989	1.239	1.690	2.167	2.833	12.017	14.067	16.013	18.475	20.278	24.321
8	0.857	1.344	1.647	2.180	2.733	3.490	13.362	15.507	17.535	20.090	21.955	26.124
9	1.152	1.735	2.088	2.700	3.325	4.168	14.684	16.919	19.023	21.666	23.589	27.877
10	1.479	2.156	2.558	3.247	3.940	4.865	15.987	18.307	20.483	23.209	25.188	29.588
11	1.834	2.603	3.053	3.816	4.575	5.578	17.275	19.675	21.920	24.725	26.757	31.264
12	2.214	3.074	3.571	4.404	5.226	6.304	18.549	21.026	23.337	26.217	28.300	32.909
13	2.617	3.565	4.107	5.009	5.892	7.041	19.812	22.362	24.736	27.688	29.819	34.527
14	3.041	4.075	4.660	5.629	6.571	7.790	21.064	23.685	26.119	29.141	31.319	36.124
15	3.483	4.601	5.229	6.262	7.261	8.547	22.307	24.996	27.488	30.578	32.801	37.698
16	3.942	5.142	5.812	6.908	7.962	9.312	23.542	26.296	28.845	32.000	34.267	39.252
17	4.416	5.697	6.408	7.564	8.672	10.085	24.769	27.587	30.191	33.409	35.718	40.791
18	4.905	6.265	7.015	8.231	9.390	10.865	25.989	28.869	31.526	34.805	37.156	42.312
19	5.407	6.844	7.633	8.907	10.117	11.651	27.204	30.144	32.852	36.191	38.582	43.819
20	5.921	7.434	8.260	9.591	10.851	12.443	28.412	31.410	34.170	37.566	39.997	45.314
21	6.447	8.034	8.897	10.283	11.591	13.240	29.615	32.671	35.479	38.932	41.401	46.796
22	6.983	8.643	9.542	10.982	12.338	14.041	30.813	33.924	36.781	40.289	42.796	48.268
23	7.529	9.260	10.196	11.689	13.091	14.848	32.007	35.172	38.076	41.638	44.181	49.728
24	8.085	9.886	10.856	12.401	13.848	15.659	33.196	36.415	39.364	42.980	45.558	51.179
25	8.649	10.520	11.524	13.120	14.611	16.473	34.382	37.652	40.646	44.314	46.928	52.619
26	9.222	11.160	12.198	13.844	15.379	17.292	35.563	38.885	41.923	45.642	48.290	54.051
27	9.803	11.808	12.878	14.573	16.151	18.114	36.741	40.113	43.195	46.963	49.645	55.475
28	10.391	12.461	13.565	15.308	16.928	18.939	37.916	41.337	44.461	48.278	50.994	56.892
29	10.986	13.121	14.256	16.047	17.708	19.768	39.087	42.557	45.722	49.588	52.335	58.301
30	11.588	13.787	14.953	16.791	18.493	20.599	40.256	43.773	46.979	50.892	53.672	59.702
35	14.688	17.192	18.509	20.569	22.465	24.797	46.059	49.802	53.203	57.342	60.275	66.619
40	17.917	20.707	22.164	24.433	26.509	29.051	51.805	55.758	59.342	63.691	66.766	73.403
50	24.674	27.991	29.707	32.357	34.764	37.689	63.167	67.505	71.420	76.154	79.490	86.660
60	31.738	35.534	37.485	40.482	43.188	46.459	74.397	79.082	83.298	88.379	91.952	99.608
70	39.036	43.275	45.442	48.758	51.739	55.329	85.527	90.531	95.023	100.43	104.21	112.32
80	46.520	51.172	53.540	57.153	60.391	64.278	96.578	101.88	106.63	112.33	116.32	124.84
90	54.156	59.196	61.754	65.647	69.126	73.291	107.57	113.15	118.14	124.12	128.30	137.21
100	61.918	67.328	70.065	74.222	77.929	82.358	118.50	124.34	129.56	135.81	140.17	149.45
200	143.84	152.24	156.43	162.73	168.28	174.84	226.02	233.99	241.06	249.45	255.26	267.54

www.ingramcontent.com/pod-product-compliance
Lightning Source LLC
Chambersburg PA
CBHW080642180526
45168CB00008B/3272
9781511820172